CAMBRIDGE EARTH SCIENCE SERIES
Editors:
A. H. Cook, W. B. Harland, N. F. Hughes,
A. Putnis, J. G. Sclater and M. R. A. Thomson

Atlas of
continental displacement

In this series

Atlas of continental displacement

200 million years to the present

H. G. OWEN

Department of Palaeontology
British Museum (Natural History)
London

CAMBRIDGE UNIVERSITY PRESS

Cambridge

London   New York   New Rochelle

Melbourne   Sydney

Published by the Press Syndicate of the University of Cambridge
The Pitt Building, Trumpington Street, Cambridge CB2 1RP
32 East 57th Street, New York, NY 10022, USA
296 Beaconsfield Parade, Middle Park, Melbourne 3206, Australia

First published 1983

Printed in Great Britain at the University Press, Cambridge

Library of Congress catalogue card number: 83-67083

British Library cataloguing in publication data
Owen, H. G.
Atlas of continental displacement, 200 million years to the present
1. Continental drift – Maps
I. Title
912'.1551136   G1046.C55

ISBN 0 521 25817 0

Contents

Contents

Preface

This volume is intended as the first of a two-part work designed to provide maps of the distribution of continental and oceanic crust during the last 700 million years of the Earth's history from the late Pre-Cambrian to the present day. It is, however, only for the last 200 million years that ocean-floor spreading information is available to indicate the chronology and mode of ocean basin development and accompanying continental displacement. When this information is plotted onto maps which assume that the Earth has been of modern dimensions throughout the last 200 million years, spherical triangular gaps (gores) appear progressively back in time from the present day. The fit of the continents together in a single supercontinent, the Pangaea recognized by Alfred Wegener, lasted until the middle Jurassic, albeit that rift valley formation had occurred earlier. However, it is only in the central 'hub' of the refit, that is, the fit of the north-west African margin into the American East Coast embayment, that the fit is perfect on a globe of modern dimensions. Radiating from this hub are gores separating regions known on geological evidence to have been in direct contact with each other at this time.

Some decades before the general acceptance of the continental displacement hypothesis, a few workers such as Hilgenberg (1933), Jordan (e.g. 1966), Egyed (1957) and Halm (1935) had speculated that such displacement might have occurred upon an Earth which was expanding its dimensions. Professor Warren Carey, during a symposium on 'continental drift' held in Hobart, Tasmania, in 1956 (Carey 1958) demonstrated that the fit of the continents together was much improved if the Earth's diameter was less than its modern value at the time of Pangaea. He referred to the earlier work on this notion carried out in Germany by Otto Hilgenberg, who published a series of reconstructions in 1933. However, neither Carey nor Hilgenberg had the benefit of ocean-floor spreading data to test their ideas and, indeed, Carey was fighting for the recognition of continental displacement at a time when most geologists considered the idea to be absurd.

The discovery that the oldest oceanic crust in the World's oceans was not older than the middle Jurassic, led Carey (1970, 1975, 1976) to advocate that all Earth expansion had occurred since then. This has become known as the 'fast expansion hypothesis' and is not supported by the available data on crustal development. By 1970, however, the so-called revolution in the Earth sciences was well under way with the general acceptance of the hypotheses of continental displacement and the development of rigid oceanic crustal plates. The latter hypothesis, together with contemporary – and current – palaeomagnetic theory, precluded the possibility of Earth expansion during Phanerozoic time.

In 1976, I presented a spherical geometric analysis of the bulk of the ocean-floor spreading evidence made available up to 1974. During this task, it was found that the continents would only fit together to form Pangaea, according to the geological evidence, when the Earth's diameter was 80% of its modern mean value. Below that figure, Pangaea could not be reformed without intra-continental dislocations. Above that figure, gores appeared in the reconstructions. Pangaea existed as a complete supercontinent until the middle Jurassic when it commenced to break up. The subsequent ocean-floor spreading patterns in the passive-margined oceans, in which the full history of continental splitting and subsequent displacement of continents apart is preserved, was found to support a near-linear increase in diameter up to the present day, consistent with the nearly straight limb of an exponential curve of increasing diameter. Despite its firm base in field data, this 'slow expansion hypothesis' is widely discounted by many geologists and geophysicists at present, although by perhaps fewer than in 1976.

A substantial increase in the amount and geographical coverage of ocean-floor spreading data since those previously analysed (Owen 1976) has prompted a new analysis of the spherical geometric implications. At the same time, the opportunity has been taken to plot the data onto maps which assume an Earth of constant modern dimensions throughout the same period of time. The conventional cartographic processes take time and the data used are those made available up to June 1980. The present atlas provides, therefore, two series of maps. The first series assumes an Earth of constant modern dimensions, while the second series assumes an Earth expanding from a diameter of 80% of its modern mean value 180–200 million years ago to its modern size. The atlas provides, therefore, a test of the validity of the two conflicting hypotheses on maps which can be tested for cartographic integrity. This is the first time that such a test has been attempted.

The second intention of this atlas is to provide base maps to facilitate studies in palaeogeography, palaeoclimatology and palaeogeophysics. In order to assist in the plotting of such data, the outline of the Earth's modern continental geography is superimposed on the reconstructions of the past. However, the actual distribution of land and sea was very different in the past. The maps given here merely represent the distributions at selected points in time of the crustal units, both continental (sialic) and oceanic (simatic), which formed the complete crust at the time of the reconstruction.

Throughout history, cartographers of the modern Earth have had to rely on numerous travellers and surveyors to provide the basic data for their maps. This atlas is no exception to that rule. The geophysical data used here are those made available up to June 1980 by the efforts of the people given in the list of references to sources and their team colleagues. However, the projection of the maps and their degree of accuracy, together with the plotting of the data, are purely the responsibility of the author.

My personal thanks are due to Mr W. B. Harland, Gonville and Caius College, Cambridge, whose original suggestion and encouragement has resulted in this atlas; to Dr H. W. Ball and Dr C. G. Adams of the Department of Palaeontology, British Museum (Natural History), without whose fundamental support this volume could not have been completed in such a relatively short period of time; and to others of my colleagues in the British Museum (Natural History) and in particular Dr G. F. Elliott, Dr M. K. Howarth and Dr R. P. S. Jefferies, for their encouragement.

Part 1
THE TEXT

Introduction

The so-called 'revolution in the Earth sciences' which has occurred during the last two decades, has been well documented in numerous books and papers. In its centenary year, the British Museum (Natural History) in conjunction with the Cambridge University Press published a two-volume work under the general title *Chance, Change and Challenge*. The first volume, called *The Evolving Earth* (Cocks, L. R. M. (Ed.) 1981), contains a series of essays on various aspects of the geological evolution of the Earth. Some of these essays review many of the present ideas and hypotheses concerning the development of the World's ocean basins and associated displacement of the continental (sialic) crustal masses. I would recommend *The Evolving Earth* as background reading for the non-specialist.

It is readily apparent from the various essay chapters in *The Evolving Earth* that the bulk of current scientific opinion favours the concept of an Earth which has possessed its modern dimensions throughout much of geological time. However, in one chapter, I have questioned whether the ocean-floor spreading evidence, which indicates how the ocean basins have developed and during what periods of time, actually supports the concept of a constant modern dimensions Earth. Although outline maps are used to illustrate the discussion, they are too small to have plotted on them the detailed ocean-floor spreading data now to hand.

In 1976, I discussed the probability of global expansion on the basis of the geological and geophysical data made available up to the early part of 1974 (Owen 1976). The field evidence to support that hypothesis could be divided into two principal categories; the first being the evidence of geological fit together at now separated continental margins, and the second being the spherical geometric implications of the growth of the ocean-floor spreading patterns.

The evidence of geological fit at continental margins had led Carey to consider the possibility of global expansion at a time when no ocean-floor spreading data were available to determine the detailed history of the splitting apart of the continents formerly grouped together as Pangaea (Carey 1958). His views were influenced by the work of Hilgenberg (1933) who, on theoretical grounds, considered that the Earth's conti-nental crust had once formed a continuous sialic shell at the lithosphere surface of an Earth some 55% of its modern diameter. This idea was speculated upon during the 1960s by workers such as Barnett (1962) and Creer (1965) and by others since. When ocean-floor spreading patterns showed that the ocean basins, including the Pacific, were not older than middle Jurassic, Carey took the extreme view that all global expansion had taken place since then and that no subduction had occurred at the Pacific margins, a requirement of the constant modern dimensions hypothesis (Carey 1970, 1975, 1976). However, such an interpretation of the field evidence requires that the Earth was shaped like a rugby football at the time of Pangaea.

In 'passive-margined' oceans such as the Atlantic, Arctic and much of the Indian, a full history of development from the initial, tensional, splitting of the continents up to the present day, is preserved. A critical examination of the evidence of fit at the common margins of these oceans indicated that the diameter of the Earth at the time of Pangaea, immediately before its break-up, was 80% of its modern value (Owen 1976). This corresponded with a short interval of time between 180 and 200 Ma which includes the late Triassic and the lower Jurassic. At diameters above the value of 80%, the continents will not fit together according to the geological data, spherical triangular gaps ('gores') appearing progressively in extent away from the centre of re-assembly of Pangaea as the diameter is increased. The fit together of the continental margins and of subsequent isochronous regions of ocean floor is affected by changing values of surface curvature. This can be illustrated by the fits of South America against Africa on two curved surfaces and one flat surface shown in figure 1 and in detail on maps 24–33. If one reduces the diameter of the Earth below a value 80% of its modern length, the continents will not fit together without increasing intra-continental displacement along major wrench fault zones which, as it happens, exist.

The analysis of the ocean-floor spreading patterns attempted in my 1976 paper was based on the passive-margined oceans, the area of which is sufficient to permit the spherical geometry of a globe and its dimensions to be determined. The spreading patterns indicate that a near-linear increase of the Earth's diameter has occurred during the last 200 million years of its history. If the sialic (continental) crust of the modern Earth once formed a complete outer crustal shell, the diameter of the Earth in the late Proterozoic would have been 55% of its modern mean value, which is not consistent with an exponential expansion during the last 700 million years (figure 2). However, there has been a substantial increment of continental crust since the Proterozoic, particularly within the Palaeozoic, which could account for part of the discrepancy.

The reconstructions given by the author in 1976 precluded the fast expansion concept advocated at that time by Carey. Moreover, the spherical geometry of the Earth at the time of Pangaea, determined by the geological evidence, showed the presence of a substantial area of earlier (Palaeozoic) oceanic crust called the Eo-Pacific. None of this crust is present today and it is logical to assume that it has been subducted. The geological evidence from the various Pacific marginal orogens, together with the ocean-floor spreading patterns within the area of the Pacific itself, indicate that marginal subduction zones were active throughout the Mesozoic and Cenozoic, in direct contradiction of Carey's view that no subduction had occurred.

In 1976, I did not provide corresponding maps of stages of continental displacement which assumed a constant modern dimensions Earth to accompany those which assumed an expanding Earth. This was a mistake, rectified here, and the opportunity was missed to provide a test of the two conflicting hypotheses at an earlier date. Predictably, the 1976 paper was the subject of substantial criticism, having transgressed elements of current geophysical theory. However, a detailed rebuttal of it has not yet been attempted on a global scale using the available ocean-floor spreading data. A number of reconstructions of the development of individual ocean basins have been published which assume a constant dimensions Earth. Most assess the field data objectively, discussing, or at least recording, the inconsistencies of fit which occur within the region concerned. Some transfer the problems of fit to regions adjacent to that described, while a few others have produced fits which are, apparently, convincing but which are, in reality, artefacts of faulty map projection. Examples of these are described in 'Some errors in reconstructions' below.

Since 1974, a substantial amount of new magnetic traverse information has become available. There are now few oceanic areas of which the basic history of development is not known and there are large areas in

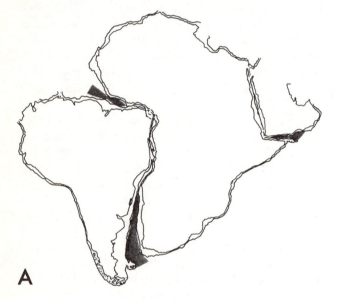

Figure 1. *The effect of changing values of surface curvature on the fit of the continents together. (A) South America and Africa with Arabia are projected separately using an azimuthal equidistant projection and then fitted together on a flat surface. Although distortions are inevitable in such a practice, the demonstration of marked V-shaped gores (shown shaded) is valid. (B) The fit of South America and Africa together, assuming an Earth of modern dimensions (diameter = 12752 km); azimuthal equidistant projection with pole indicated ∗. The V-shaped gores are greatly reduced in area, but are still present. (C) The fit of the continents together assuming an Earth with a diameter 80% of its modern mean value (10202 km); azimuthal equidistant projection with pole indicated ∗.*

A

B

C

which the growth patterns are known in some detail. The development of certain areas of high deformation such as the Mediterranean region, the Alpine–Himalaya belt, South East Asia and the Philippines and the West Antarctic Peninsula, remain to be resolved in detail. In this atlas, the reconstructions which assume a constant modern dimensions Earth retain the modern configuration of these deformed areas, to conform with similar series published by other authors (e.g. Smith, Hurley & Briden 1981, see Appendix Note 1). In the reconstructions assuming an expanding Earth, the broad outline of regional deformation which is consistent with the adjacent ocean-floor spreading evidence is given, but it is grossly over-simplified. The fact that palaeomagnetic evidence indicates that certain zones in Japan were widely separated along a major wrench fault zone at a particular time, while they are shown here in their modern relationship, does not invalidate the reconstructions based on the ocean-floor spreading data. It merely reflects that the author has concentrated his efforts and capacity in assimilating the spreading data and assessing its spherical geometric implications and that his capacity to include detailed additional data is limited.

This Atlas provides a test of the ocean-floor spreading information on two conflicting models; one which assumes that the Earth's dimensions have been constant during the last 200 million years, the other which assumes an increase in diameter from 80% of its modern mean value 180–200 million years ago to its present value. This is the first attempt at such a test of the basic field data upon which the determination of the mode and chronology of ocean basin development depends.

Data sources, handling methods and limits

All global cartographers rely on the work of countless surveyors to provide the basic information from which their maps are constructed. This is true equally of the great map-makers of the past, such as Ptolomy, Contarini, Mercator and Jodocus Hondius among many others, right up to the present day. Today, the primary surveyor is being partly replaced by the optical/electronic systems in high-flying aircraft and orbital satellites. Even the conventional cartographer might one day be replaced by the skillful use of the computer (e.g. Monmonier 1982).

The present atlas is no exception to this rule. The basic data have been collected by numerous individuals operating ship-borne and air-borne towed magnetometers, supported by crews keeping vessels and aircraft on accurately positioned traverses. Others have interpreted the magnetometer traces and checked the data against the information obtained from deep ocean borehole core sequences. The result of all this effort and expense, is a widespread coverage of magnetic anomaly data over the World's oceanic crust which, in some areas, is very detailed. If one adds to this the wealth of geological and geophysical survey information from the continental margins, a very good picture of the development and age of the ocean basins can be obtained (e.g. Nairn, Stehli *et al.* (Eds.) 1973–81).

The ocean-floor spreading data used in this atlas are those made available up to July 1980. Some of the more recent principal papers published after the preparation of the global reconstructions mapped here, are referred to in the Appendix. In order that the plotting of the information may be checked, it is shown here on three maps employing the conventional Mercator's projection (Maps 1–3). Map 1 shows the North and South Atlantic Oceans, the spreading data being derived from the following sources: Barker (1970, 1972a, b), Barrett & Keen (1976), Bergh (1977), Bergh & Barrett (1980), Cande & Kristoffersen (1977), Dickson, Pitman & Heirtzler (1968), Hayes & Rabinowitz (1975), Herron & Tucholke (1976), Johnson & Vogt (1973), Keen, Hall & Sullivan (1977), Kristoffersen (1978), Kristoffersen & Talwani (1977), Kumar & Embley (1977), La Brecque & Hayes (1979), Ladd, Dickson & Pitman (1973) Larson & Hilde (1975), Larson & Ladd (1973), Larson & Pitman (1972), Lattimore, Rona & De Wald (1974), Laughton (1971, 1972), Le Pichon & Fox (1971), Le Pichon & Hayes (1971), Mascle & Phillips (1972), Olivet, Le Pichon, Monti & Sichler (1974), Peter, Lattimore, De Wald & Merrill

Figure 2. Exponential curve of the value of the Earth's mean diameter through time, assuming today's value, a value of 80% of the modern diameter 180–200 Ma B.P., and the amount of sialic crust known, or thought, to have been formed by the late Proterozoic, which it is assumed formed a complete sialic shell. The curve also assumes that the Earth has retained its shape as a sphere of rotation throughout this period, and that the radioactive decay rate used in the dating has been constant.

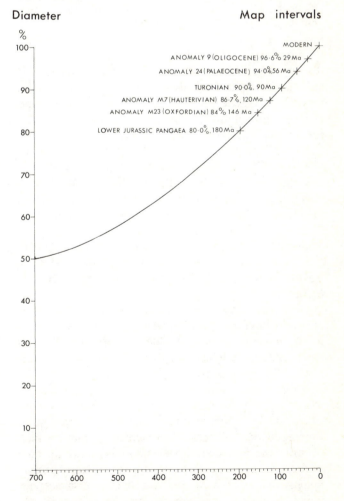

Geological age in millions of years (Ma)

(1973), Phillips, Fleming, Feden, King & Perry (1975), Pitman, Larson & Herron (1974), Pitman, Talwani & Heirtzler (1971), Purdy & Rohr (1979), Rabinowitz, Cande & Hayes (1979), Rabinowitz & La Brecque (1979), Rabinowitz & Purdy (1976), Ramberg, Gray & Raynolds (1977), Sclater, Bowin, Hey, Hoskins, Peirce, Phillips & Tapscott (1976), Steiner (1977), Storetvedt (1972), Talwani & Eldholm (1977), Taylor & Greenwalt (1976), Van Andel, Rea, Von Herzen & Hoskins (1973), Vogt, Anderson & Bracey (1971), Vogt & Avery (1974), Vogt & Einwich (1979), Vogt & Johnson (1971), Vogt & Ostenso (1970), Williams & McKenzie (1971). Map 2 shows the Indian Ocean, the spreading data being derived from the following sources: Bergh (1977), Bergh & Norton (1976), Bowin, Purdy, Johnston, Shor, Lawver, Hartano & Jezek (1980), Hayes (1972), Hayes & Ringis (1973), Heirtzler, Cameron, Cook, Powell, Roeser, Sukardi & Veevers (1978), Larson (1975, 1977), Larson, Carpenter & Diebold (1978), McKenzie & Sclater (1971), Markl (1974, 1978), Norton & Sclater (1979), Schlich (1974), Sclater & Fisher (1974), Sclater, Luyendyk & Meinke (1976), Ségoufin (1978), Simpson in Norton & Sclater (1979), Weissel & Hayes (1972). Map 3 shows the Pacific Ocean, the spreading data being derived from the following sources: Anderson, Clague, Klitgord, Marshall & Nishimori (1975), Anderson, Moore, Schilt, Cardwell, Tréhu & Vacquier (1976), Atwater & Menard (1970), Ben-Avraham, Bowin & Segawa (1972), Ben-Avraham & Uyeda (1973), Bowin, Purdy, Johnston, Shor, Lawver, Hartano & Jezek (1980), Bracey (1975), Christoffel & Falconer (1972), Christoffel & Ross (1970), Cooper, Scholl & Marlow (1976), Elvers, Srivastava, Potter, Morley & Sdidel (1973), Falconer (1972), Handschumacher (1976), Hayes & Pitman (1970), Hayes & Ringis (1973), Hayes & Taylor (1978), Herron (1971, 1972), Herron & Tucholke (1976), Hey (1977), Hey, Johnson & Lowrie (1977), Hilde, Isezaki & Wageman (1976), Hussong, Wipperman & Kroenke (1979), Kobayashi & Isezaki (1976), Larson & Chase (1972), Larson & Pitman (1972), Lonsdale & Klitgord (1978), Louden (1976, 1977), Luyendyk, Bryan & Jezek (1974), Luyendyk, MacDonald & Bryan (1973), Malahoff & Handschumacher (1971), Mammerickx, Anderson, Menard & Smith (1975), Molnar, Atwater, Mammerickx & Smith (1975), Murakami, Tamaki & Nishimura (1977), Sclater & Klitgord (1973), Tamaki, Joshima & Larson (1979), Truchan & Larson (1973), Vogt & Byerly (1976), Vogt & De Boer (1976), Watts & Weissel (1975), Weissel & Hayes (1972, 1977), Weissel & Watts (1975, 1979).

The Arctic Ocean cannot be displayed using Mercator's projection and the following additional sources are used in the modern Arctic map (Map 4): Jackson, Keen & Falconer (1979), Johnson & Vogt (1973), Vogt, Taylor, Kovacs & Johnson (1979).

Conventional cartographic methods have been employed in the projection of the maps given in this atlas. Although the computer is used extensively nowadays in the production of ocean-floor spreading maps which illustrate the data, it has been employed successfully only in those maps which portray the modern Earth. In map reconstructions of stages of continental displacement which assume a constant modern dimensions Earth,

errors are apparent in the projections which indicate imperfect programming. The employment of the computer in the production of detailed ocean-floor spreading maps which assume an expanding Earth, is not a time-effective proposition at present, in terms of programme writing and the costly employment of the software and hardware facilities.

The handling of the data using conventional cartographic methods is quite simple. The information is plotted onto individual grid units with dimensions of 10° of latitude and 10° of longitude coinciding with the Earth's conventional co-ordinate net. Each unit is projected separately using an azimuthal equidistant projection with the pole situated at the centre of each unit. The amount of distortion at the margins of each unit assuming a scale globe of 380 mm diameter is negligible. These units can be used to transfer data to the surface of a globe of any given dimensions in order to test the fit of crustal units together, or they can be used as the basis of maps employing the equidistant projection. The method also lends itself to easy conversion to computer raster data, once accurate programmes have been developed. Because one is dealing with segments of almost true scale surface area, it is possible to build up a mosaic accounting for all the surface area of the Earth at a given point in time. However, once this mosaic is completed it is necessary to transform the reconstruction to a readable map form by re-projection as described in the 'Notes on the cartographic projections' below.

The limits of accuracy in data handling fall into three main categories. In the first instance one can ask the questions; how accurate are the magnetometer records and the course navigation; are the patterns of course tracks sufficient to indicate trends and are the profiles correctly interpreted in terms of the vectors of the anomaly lineations and of their dating? It so happens that when a critical examination of the spreading patterns is made, such as in this atlas, remarkably few problems have arisen which could be put down to major technical or interpretive errors, although some exist. In so far as the dating of magnetic anomalies is concerned (Table 1), it is desirable to have corroborative Deep Sea Drilling Project (DSDP) borehole information at salient points in order to check the magnetic anomaly sequence. Obviously, there are geographical and economical limits to such drilling. At the scale employed here, most of the errors are not of sufficient magnitude to affect the reconstructions.

The second category of accuracy in handling the data concerns the theoretical basis upon which the reconstructions are made. Can we be sure that there is no form of discrete crustal subduction in passive-margined oceans which has not yet been detected? Even had this occurred, it would not explain the gores which are a feature of the constant modern dimensions Earth reconstructions. The assumption of the 1000 m isobath as the effective edge of the continental crust has been questioned by some (e.g. Hallam 1976). However, in passive-margined oceans, the position of the crustal transition zone (the zone between deep-faulted continental crust and oceanic crust) and the commencement of the magnetic anomaly sequences *can be measured from the existing modern coastline*. The argument whether or not the 1000 m isobath represents the edge of the continents world-wide, is irrelevant even in the reconstructions of Pangaea. Apart from facilitating the reading of the maps showing the dotted modern coastline as a guide, the practice allows, also, a check to be made on the accuracy of the plotting of the oceanic crustal information in the reconstructions.

The third limit of accuracy involves the development of certain relatively small, but important, regions for which there was no spreading data available at the commencement of map construction. These regions include the Amerasia Basin of the Arctic Ocean, the Weddell Sea off Antarctica and some of the west Pacific marginal basins. The areal limits in which their development could have occurred, can be determined accurately, however, from the surrounding oceanic regions in which the spreading patterns are known. Nonetheless, the detailed pattern of their development and the chronology, remains conjectural. The reconstructions of the development of these regions given here are possible, but they will have to be revised when definitive

Table 1. *Magnetic polarity time scale for the Mesozoic and Cenozoic*

Anomaly no.	Estimated age Ma B.P.*	System and stage	
1	0.0–0.7	QUATERNARY	Pleistocene
2	1.6–1.8		
2A	2.4–3.3	TERTIARY (Neogene)	Pliocene
3	3.7–4.6		
3A	5.1–5.6		Miocene
4	6.4–7.0		
5	8.3–9.7		
5A	10.9–11.5		
5B	14.3–14.7		
5C	15.7–16.5		
5D	17.1–17.7		
5E	18.1–18.7		
6	19.0–20.0		
6A	20.5–21.4		
6B	22.2–22.6		
6C	23.0–23.9	(Palaeogene)	Oligocene
7	25.2–25.7		
7A	26.1–26.3		
8	26.6–27.5		
9	28.0–29.0		
10	29.6–30.2		
11	31.1–32.0		
12	32.4–32.8		
13	35.3–35.9		
15	37.3–37.7		
16	38.1–39.3		Eocene
17	39.6–41.2		
18	41.4–42.9		
19	43.8–44.2		
20	44.9–46.4		
21	49.0–50.7		
22	52.3–53.0		
23	54.3–55.1		
24	55.6–56.6		Palaeocene
25	58.7–59.2		
26	60.0–60.4		
27	62.3–62.7		
28	63.3–64.0		
29	64.3–64.9		
30	65.4–66.8	CRETACEOUS (upper)	Maastrichtian
31	66.8–67.6		
32	69.2–71.0		
33	71.6–76.5		Campanian (part)
34	79.6–87.5		Santonian
unnumbered (reversed)†	87.5–88.0		Turonian
35 & 36†	88.0–103.8		Turonian–Albian
unnumbered (reversed)†	103.8–104.0	(lower)	Albian
M0‡	110.9–111.6		Aptian
M1	114.3–114.7		Barremian
M2 (normal)	114.7–115.4		
M3	115.4–117.5		
M4 (normal)	117.5–118.5		
M5	118.5–119.0		Hauterivian
M6	119.2–119.3		
M7	119.5–119.9		
M8	120.2–120.4		
M9	120.7–121.2		
M10	121.5–121.9		
M10N	122.9–123.2		
M11	124.5–124.8		Valanginian
M12	126.0–126.7		
M13	127.7–128.1		
M14	128.3–129.2		
M15	129.8–130.3		
M16	132.1–132.8		Berriasian
M17	133.3–134.9		
M18	135.5–136.0	JURASSIC (upper)	Tithonian
M19	137.6–137.9		
M20	139.0–139.9		Kimmeridgian
M21	141.2–141.7		
M22	143.7–144.6		
M23 (normal)	145.6–145.7		Oxfordian
M24	147.2–147.4		
M25	148.7–149.0		
	150.0§		Callovian

* Estimated interval in millions of years before present (B.P.) to the nearest 100 000 years.

† These include the normal and reversed anomalies mapped in the southern Pacific in the region of New Zealand, and those labelled A–C in the south-western part of the Wharton Basin of the Indian Ocean.

‡ The age calibration of the M series of magnetic anomalies given by Larson & Hilde (1975) has been revised by Vogt & Einwich (1979). However, the ammonite evidence available agrees more closely with the Stage correlations of Larson & Hilde rather than those of Vogt & Einwich. The arguments used by Vogt & Einwich in their revision of Larson & Hilde, illustrate the amount of uncertainty that there is in the dating of magnetic anomalies at fine detail level. This variation does not affect, fundamentally, the reconstructions given in this atlas.

§ Long periods of normal polarity with little reversal activity occur in the Cretaceous, between the Aptian and the Coniacian, and in the lower and middle Jurassic. Crust generated during these intervals is described as being magnetically 'quiet'. Low-amplitude reversals have been detected, however. One example, with a revision of dates, extends the M sequence to M29 (*ca* 157 Ma) within the Callovian (Cande, Larson & La Brecque 1978).

Compiled essentially from La Brecque, Kent & Cande (1977) and, with modifications, from Vogt & Einwich (1979). Anomalies 1–34 are normally oriented although they include minor reversals. Anomalies M0–M26 are reversed unless indicated otherwise.

data are made available. Because of subduction around the Pacific margin, the exact positions of the spreading patterns in the Pacific reconstructions given here may also need to be revised a little in due course.

It remains to make the obvious point concerning the limits of accuracy. That is, has the author made the reconstructions accurately? The maps provided in this atlas can be checked for any such error.

Description of the reference co-ordinate graticule

Most reconstructions of continental displacement on a world-wide scale, have the latitude–longitude co-ordinate graticule oriented according to the geomagnetic field data for the time-period concerned. This tends to detract from a clear picture of the motions of continental crustal units relative to each other in the Palaeozoic, because of the marked effect of polar wandering. Even in the Mesozoic and Cenozoic, where polar wandering is not marked, it is much better to use a common graticule to which all motions of continental crust and the development of ocean basins can be referred.

In order to determine crustal unit motions and growth on an Earth of constant modern dimensions throughout Phanerozoic time, the orientation of the Earth's modern co-ordinate graticule is taken to be fixed in space. That is, the positions of the north and south geographic poles and lines of latitude and longitude in each reconstruction which assumes a modern dimensions Earth, coincide in exact position with the Earth's modern co-ordinate graticule. To determine the position of a continental unit, or an area of oceanic floor, relative to this reference graticule, the pattern of ocean-floor spreading is used as far as possible. Thus, for example, as the North Atlantic Ocean develops by ocean-floor spreading in a nearly symmetrical manner each side of the spreading axis (the mid-Atlantic Ridge), North America and Europe are displaced away from each other to arrive, eventually, at their modern co-ordinate positions. Obviously, by following the procedure in reverse, one can determine the relative co-ordinate positions of any crustal unit at any point in time back to Pangaea (180–200 Ma) by reference to the spreading patterns. This method of determining crustal positions in the past indicates relative motion and not absolute motion in terms of the whole Earth. It is not known at present whether the various shells of the Earth, such as the outer core, lower and upper Mantle, rotate together relative to the inner core, or differentially to each other.

The determination of former co-ordinate positions on a constant modern dimensions Earth described above, is straightforward, albeit that spherical triangular gaps (gores) develop progressively as one goes back in time towards Pangaea. However, in the case of reconstructions which assume an expanding Earth, a somewhat different approach is required. It is necessary in the first instance, to determine the value of surface curvature (and thus the diameter of the Earth) at which the ocean-floor spreading patterns in passive-margined oceans, up to a selected pair of magnetic anomalies, will fit together without overlap or gaps. Exactly the same principle applies to the

determination of the size of the Earth at the time of the re-assembly of the continents to form Pangaea, except that the evidence of fit is provided by truncated structures on the corresponding continental margins. The spherical geometry can only be determined in passive-margined oceans, in which a full history of continental splitting and subsequent oceanic crustal separation, is preserved. The regions of the Arctic, North and South Atlantic and the bulk of the Indian Ocean taken together, are sufficient to determine the size of the sphere in a given reconstruction.

The method of reconstruction is as follows. The spreading data forward in time to a selected pair of magnetic anomalies in the various passive-margined ocean basins, are plotted on a sphere of which the diameter can be varied in order to determine the best fit together in response to changes in surface curvature (Owen 1976). In practice, I have determined these fits at approximately 30 million year intervals back to the re-assembly of Pangaea. This is a lengthy but unavoidable process, which has to be carried out as carefully as possible. At the end of such an investigation, one has a series of complete crustal shells of decreasing surface area at time intervals of approximately 30 million years back to 180–200 Ma on which the continental displacement history of the Mesozoic and Cenozoic Earth has been determined.

The second step in the process is the orientation of these shells to a fixed co-ordinate graticule in order to demonstrate relative motion. We can ask the same question as in the case of the constant dimensions Earth reconstructions; what are the co-ordinate positions of the continents re-assembled in Pangaea 180–200 million years ago, which will be displaced to their modern co-ordinate positions in response to the growth of the ocean-floor spreading patterns?

In the expanding Earth graticules, the vector of the modern 0° meridian and the position of the north geographic pole are fixed in space. The remainder of the latitude–longitude co-ordinate net is constructed from these two positions, the size of the graticule depending on the diameter of the Earth at the time of the reconstruction determined previously. The crustal shells are then oriented to their respective graticules according to the relative displacement data as shown in the maps given here.

Notes on the cartographic projections

Having produced the crustal shells for both the constant modern dimensions and expanding Earth reconstructions and oriented them to their respective co-ordinate graticules, the next step is to produce suitable maps. The maps should be to a common scale, of sufficient size to be able to display the spreading data diagrammatically and should cover the whole of the oceanic area of the Earth with the minimum of distortion. The Earth is a sphere, a fact unwittingly forgotten by some who attempt reconstructions of continental drift. All maps and plans are flat representations of a curved surface; the surface of the Earth. Away from the pole or origin of the map, the area and configuration of geographical units will become distorted progressively. In order to permit the

Figure 3. Distortion characteristics of Mercator's projection, shown graphically in comparison with a true scale representation of a 10° longitude by 90° latitude co-ordinate spherical triangle taken from a globe.

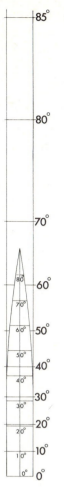

amount of distortion to be calculated, maps and plans are drawn, by convention, employing a co-ordinate graticule projected onto a flat surface from the sphere according to definite mathematical formulae. There are a large number of such projections, those most useful being found commonly employed in atlases (see Steers 1962). However, since the advent of the computer, other projections have been calculated, some of them with co-ordinate graticule patterns little short of bizarre (e.g. Tobler 1973).

Three projections are used in this atlas. Maps 1–3 employ Mercator's projection, the distortion scale of which is shown graphically in figure 3. It is the convention among oceanographical geophysicists to display ocean-floor spreading data on maps employing Mercator's projection. There are only two advantages that the writer can see in this practice. The first is that ship and aircraft tracks can often be represented by straight lines and that plotting of information is that much simpler. The other advantage is that there is more space in the higher latitudes of both hemispheres to present data which would be severely cramped on maps using other projections. However, the important spreading information from the Arctic regions cannot be displayed properly, if at all, and the relationships of the Boreal Atlantic and North Pacific and the peri-Antarctic spreading zones are rendered meaningless except to the cartographer. These factors, together with the fundamental distortion characteristics of the projection, render it unsuitable for the display of spreading data. Nonetheless, this is the projection in common use and, in order to provide an absolute check on the field data used in the reconstructions, the three maps employing Mercator's projection of the Atlantic, Indian and Pacific Oceans are included here (Maps 1–3).

The detailed reconstructions presented here employ the polar, oblique and modified equatorial cases of the azimuthal equidistant projection. The characteristics of this projection are best illustrated by the polar case. In this

Figure 4. Diagrammatic representation of the polar case (A) and the oblique case (B) of the azimuthal equidistant projection. The plane of the map should touch the surface of the Earth at right angles to the intercept of the projection axis (projection pole). For clarity, the plane of the map is raised above this point of contact.

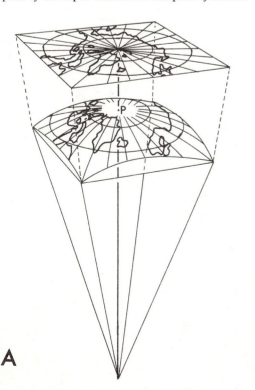

A

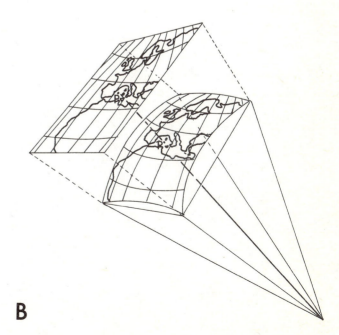

B

case, the geographic north or south pole coincides with the projection pole of the map (see figure 5). Each radian from it coincides with a line of longitude (meridian) and is of true scale length. The intercepts upon each radian of a line of latitude are at true scale distance from the pole. However, as we are constructing the co-ordinate graticule on a plane surface from a sphere (figure 4A), meridians will diverge from each other progressively, above true circumferential scale rate. The parallel lines of latitude increase progressively in length over true scale length according to the percentage scale given in Table 2. Maps 4–13 and 54–63 employ this simplest of cases.

The oblique case is used in Maps 14–43. Exactly the same radial true scale and circumferential distortion characteristics apply in this case outward from the selected projection pole. In this case, however, the lines of latitude and longitude form arcs upon the map (figure 4B) and the relationship of the two cases is shown in figure 5. Within a span of 100° of longitude in the oblique case, the amount of distortion towards the outer regions is reasonable, visually, corresponding as it does to a radius of 50° from the projection pole. This is acceptable for the

display of the growth of the North and South Atlantic and the Indian Ocean, but the Pacific requires a map of twice this longitudinal span.

The conventional equatorial case of the azimuthal equidistant projection was used extensively by the writer in 1976 and the distortion above the value of 50° from the projection pole can be assessed readily enough from those maps. To extend the projection sufficiently to cover the area of the Pacific, would make the outer regions unrecognizable. A useful modification of the equatorial case has been devised by Brigadier Guy Bomford for *The Oxford Atlas* (Lewis & Campbell (Eds.) 1951). In this modification, the conventional equidistant projection is

Figure 5. The co-ordinate graticules of the polar case (projection pole 90°N or S) and the oblique case used herein (projection pole at 22°N or S latitude), superimposed so that the projection poles are common, in order to show the relationship of the two cases; azimuthal equidistant projection. The amount of circumferential distortion shown in the polar case in % is applicable equally to the oblique case at the same radial distance from the projection pole.

Table 2. *Circumferential distortion expressed in percentages of true scale distance; azimuthal equidistant projection*

Polar case latitude°	Distortion factor %	Oblique case distance from origin°
90	0.0	0
80	+0.5	10
70	+2.1	20
60	+4.7	30
50	+8.6	40
40	+13.9	50
30	+20.9	60
20	+30.0	70
10	+41.7	80
0	+57.1	90

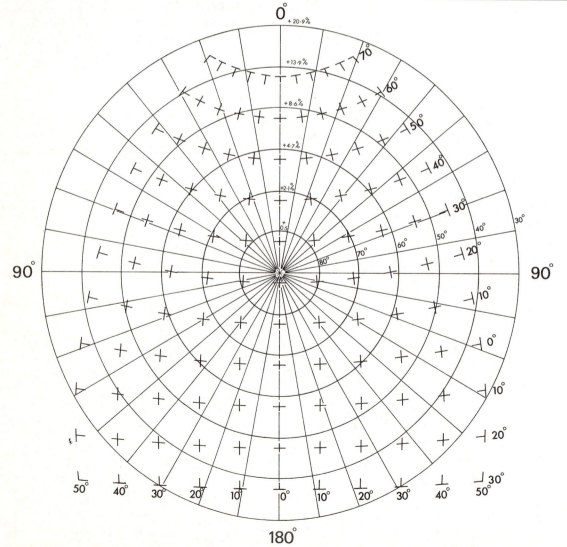

stretched so that while radial distances retain their true scale characteristic, a uniform distortion of 37% is produced around a bounding oval. This is shown graphically in figure 6, in which the pecked lines indicate the degree of bending of the radians to reduce the distortion at the west and east Pacific margins. This modified azimuthal equidistant projection is used here for the Pacific reconstructions (Maps 44–53).

Apart from the detailed maps of oceanic crustal growth, World outline maps spaced at approximately 30 million year intervals are included for palaeogeographic, palaeoclimatological and faunal distribution studies. The best class of maps for this purpose is the ellipticals and, of these, the third example described by Winkel (1921) is used here. This is an arbitrary map in the sense that the graticule is an arithmetical mean between the simple cylindrical projection (Plate Carrée) and Aitoff's projection. It is used for Maps 64–76.

Figure 6. The modification of the equatorial case of the azimuthal equidistant projection, devised by Brigadier Guy Bomford in order to reduce the distortion of geographical outlines in the outer zones of the map, and used herein for the portrayal of the north and central Pacific. The divergence of the pecked lines from their corresponding radians, indicates the amount of bending towards the ellipse, evident in the co-ordinate graticule.

Some errors in reconstructions

The bulk of ocean-floor spreading data is handled by computers. This is an excellent method of plotting data obtained from the Earth's surface and the draughting of specific maps as required. Problems have arisen, however, when reconstructions of continental displacement have been made assuming a constant modern dimensions Earth. Two examples may be cited in which the authors concerned have overlooked the fact that, if one changes the position of a crustal unit relative to the pole or origin of the projection, the unit concerned will change shape and, or, apparent area on the map. When such operations are carried out employing a computer programme, it suggests that the computer is not programmed to recalculate the effect of a change of position of a surface unit on the face of a sphere and, indeed, that the raster data mode assumes a flat surface.

The first example is the common fault of merely rotating South America in stages to fit against Africa on maps employing Mercator's projection (e.g. Mascle & Phillips 1972; Rabinowitz & La Brecque 1979). The fit together at the time of Pangaea is apparently perfect without the triangular gap evident in the constant dimensions Earth map given here (Map 32). The act of rotating South America on the map, however, implies an alteration of its position on the globe relative to the position of the projection pole or origin. Although its shape would have been a correct representation on the

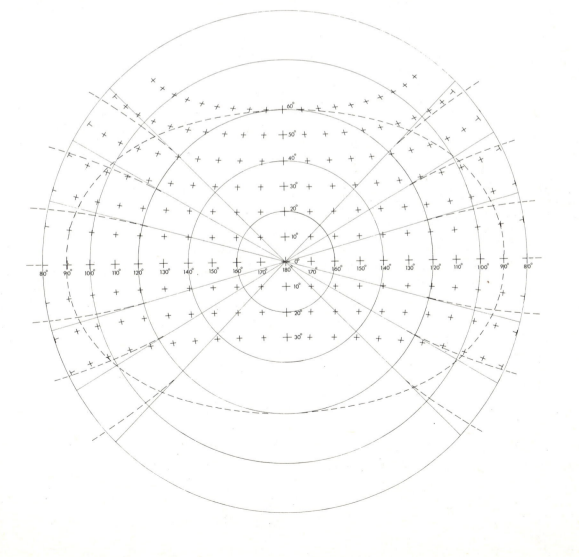

Figure 7. An example of an invalid cartographic reconstruction. (A) is the modern map of the Indian Ocean (taken from Map 34). (B) is a 'reconstruction' at Anomaly 24 (Palaeocene) made by removing all crust on the modern map generated after the dykes of Anomaly 24, and merely fitting the remainder together (co-ordinates derived from figures 14 and 16 of Norton & Sclater 1979). In effect, the chimera is superimposed upon the co-ordinate graticule of an azimuthal equidistant projection with pole situated at 22°S latitude 80°E longitude which has been correctly drawn. However, the map has no corrections for changes in circumferential distortion required as the geographical units are moved in relation to the projection pole. (C) is a correctly projected map assuming the Earth's modern dimensions and giving the nearest fit at Anomaly 24 that can be obtained (modified from Map 36).

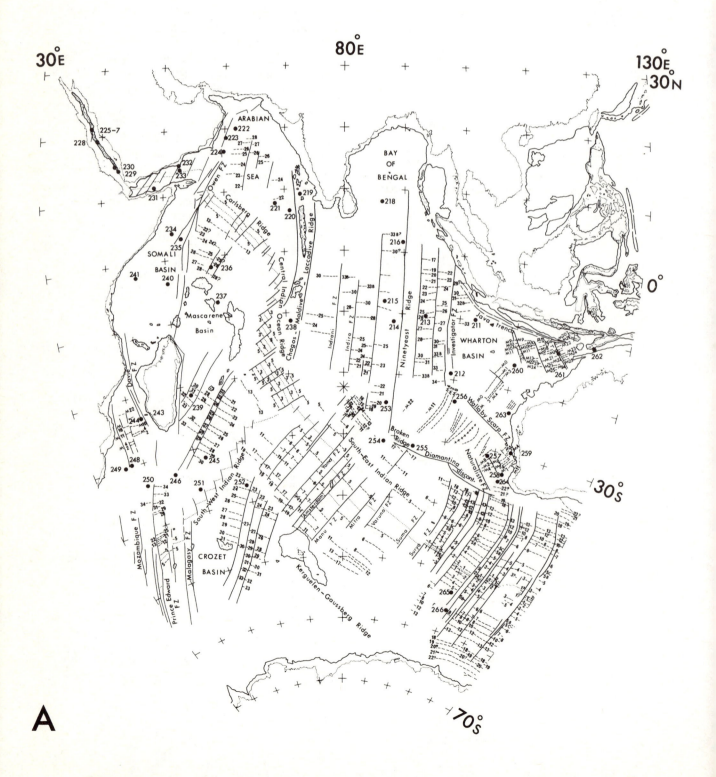

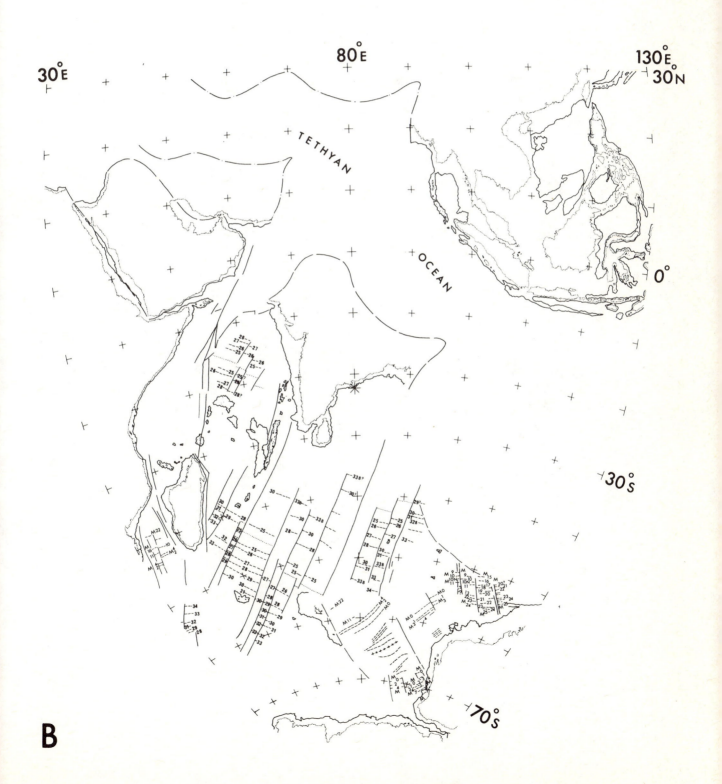

B

Figure 7. (Continued)

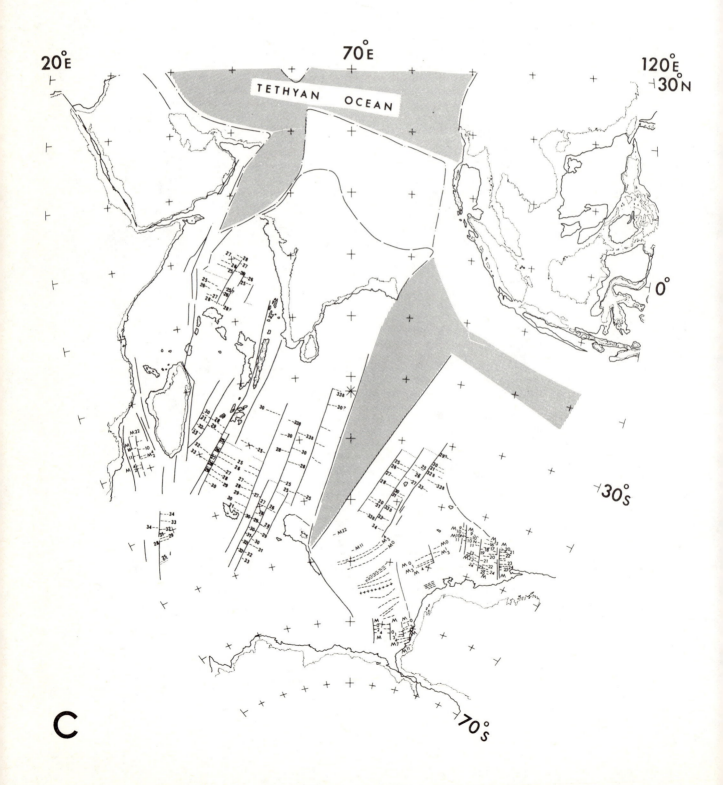

C

original map, the change of position requires that South America be re-projected to accord with its new attitude to the origin of the map projection: it will change shape. The change of shape when South America is correctly projected produces the gore seen shaded in Map 32 which widens progressively southward between South America and Africa.

A less obvious error, but fundamentally the most misleading, is to be seen in the reconstructions of the development of the Indian Ocean given by Norton & Sclater (e.g. 1979). The authors claim to use an azimuthal equal area projection which, by inspection, has a pole at about 22° south latitude, 20° east longitude. The ocean-floor spreading evidence is shown in detail on Mercator's projection and this is transferred onto their oblique azimuthal modern map. In their reconstructions, which assume a constant modern dimensions Earth throughout the last 200 million years, the spreading data appear to fit perfectly and to support, therefore, the constant dimensions Earth hypothesis.

Unfortunately, Norton & Sclater's maps are not projected on to an oblique equal area graticule, but on to an oblique equidistant graticule. The fundamental error, however, lies in their computer programme, which does not possess the function to re-project the maps as each relevant area of ocean floor is subtracted from the modern map and the remainder rotated together. Each reconstruction is an artefact best indicated by the 'reconstruction' given here in figure 7B. Figure 7A shows the modern Indian Ocean with its crustal spreading patterns plotted on it (see Map 34). If one removes all crustal area generated after Anomaly 24 *on the map* and the remainder is then brought together, a nearly perfect fit can be obtained as shown in figure 7B, in which the co-ordinate positions are derived from Norton & Sclater's figures 14 and 16. However, from the earlier discussion of the characteristics of distortion in azimuthal equidistant projections, it becomes apparent that areas of higher circumferential distortion have been moved inward towards the projection pole, to areas where the circumferential distortion is of a lower value, without any correction being made. In effect, the surface area has been increased in a non-linear manner and the map does not possess cartographic integrity. If, on the other hand, the post-Anomaly 24 spreading area is subtracted from the surface of the modern globe and the subsequent configuration on that globe of an Anomaly 24 reconstruction is re-projected, we arrive at the map shown in figure 7C with its gores readily apparent (see also Map 36).

Maps such as those discussed above, fail to comply with the fundamental laws of map projection and do not possess cartographic integrity. Many workers consider the computer to be an essential instrument for the production of maps of continental displacement. This is neither correct nor is it advisable at present. Standard cartographic procedures are straightforward and, although tedious and time consuming in their application, can produce maps of greater accuracy. It is essential that the spherical geometry indicated by the ocean-floor spreading patterns, be checked accurately in order to assess the physical implications of global expansion.

Development of oceanic regions

In the principal part of the atlas, the growth of the ocean basins and the accompanying dispersal of the continents once assembled together in Pangaea, are illustrated in detail. Two series of maps are presented in each section. In the case of the reconstructions which assume that the Earth has not changed its dimensions during the last 200 million years, the time interval between each reconstruction is approximately 60 million years. In the reconstructions which assume an expanding Earth, the intervals are reduced to approximately 30 million years.

A descriptive outline of oceanic development accompanies the maps and is divided up into the separate ocean regions. Although this method achieves clarity when dealing with the individual regions, it obscures the fact that crustal development and displacement in one area of the globe affects all other areas. Problems of fit in one area cannot be conveniently transferred to an adjacent region and then ignored, as so often happens in the scientific literature at present. A discussion of the development of the Arctic, South Atlantic and Indian Oceans written by the author, is to be found with other papers giving the case for and against global expansion in a volume marking the Expanding Earth symposium held in Sydney in February 1981 (Carey (Ed.) 1983).

A list of maps is provided in the Contents arranged under individual oceanic regions. In order to permit easy reference to contemporary maps of adjoining regions projected from the same global reconstruction, a series of index maps is given on pp. 34–37.

SECTION 1

Boreal region

MAPS 4–13

The Arctic Ocean consists, essentially, of two large basins; the Amerasia and Eurasia Basins, separated by the Lomonosov ridge (Map 4). Detailed spreading patterns are available from the Eurasia Basin and show that the bulk of its crust has been generated from the still active Nansen (or Gakel) central ridge during the Cenozoic (e.g. Vogt, Taylor, Kovacs & Johnson 1979). The Nansen ridge is a continuation of the mid-Atlantic ridge offset by the Spitzbergen fracture zone. Map 4 shows that the Eurasia Basin is, in reality, an end Cretaceous to modern extension of the North Atlantic into the Arctic region.

The development of the Amerasia Basin is less clear in the absence of sufficient magnetic traverses to elucidate the spreading patterns (e.g. Vogt, Taylor, Kovacs & Johnson 1979, but see Appendix Note 2). There are no active spreading centres today and it is apparent that spreading ceased to form the crust of this basin in the early Tertiary. The Canada Basin may have an older, late Jurassic and lower Cretaceous spreading history, rather earlier than the upper Cretaceous and early Tertiary spreading history inferred here, for the bulk of the Amerasia Basin. Despite the absence of clear spreading patterns, it is possible to produce a series of reconstructions of the growth of the Amerasia Basin and, with the pattern in the Eurasia Basin, of the Arctic Ocean as a whole. These reconstructions are consistent with the major clockwise tectonic motions of North America, relative to Eurasia and Greenland, in response to oceanic crustal spreading in the North Atlantic. The major difficulty in the reconstructions given here is in the determination of the precise chronology of the various stages of displacement prior to the late Cretaceous. The reconstructions also require an expanding Earth.

In essence, North America has rotated in a clockwise manner relative to Eurasia and Greenland since the early upper Jurassic commencement of ocean-floor spreading in the southern North Atlantic. Alaska appears to have acted as a pivot for this motion. The development of the Amerasia Basin reflects the earlier stage of the growth of the North Atlantic crust, which led to the splitting away of Canada from Greenland in the late Cretaceous and early Tertiary. This spreading trend, which produced the Labrador Sea, Davis Strait and Baffin Bay, ceased to be active in the early Tertiary (early Oligocene). In the Palaeocene, active spreading had commenced to form the crust of the Denmark Strait and the Norwegian–Greenland Sea and this has become the active spreading region which has separated Greenland from the European margin. Generation of oceanic crust from the mid-Atlantic Reykjanes–Iceland and Mohns–Knipovich ridges continues today. The Eurasia Basin of the Arctic Ocean is a continuation of this spreading phase, extending over the same Palaeocene to Recent time span and merely offset by the Spitzbergen fracture zone.

There would be little dispute now concerning the foregoing outline of the development of the Arctic Ocean basins. The fundamental cause of the evolution of the Arctic region as a whole was discussed by Carey (1958, 1976). He indicated some of the discrepancies which exist between the geological evidence obtained from the surrounding continental regions and the geometry of development of the Arctic Ocean, if one assumes a constant modern dimensions Earth since the time of Pangaea. If one makes such a series of reconstructions comparable to those of Firstbrook, Funnell, Hurley & Smith (1980) and Smith, Hurley & Briden (1981), assuming the modern configuration of the Arctic continental regions, a spherical triangular gore is produced in each case (Maps 6, 9 and 12). Each gore has its apex to the south, separating the Canadian Arctic Islands from Greenland, and widens towards the Pacific, leaving a gap between the North American and Russian continental margins. Even when the Anadyr–Verkhoyansk Block is moved northward along the line of the Orulgen–Verkhoyansk orogen into the 'Palae-Arctic Ocean', the gore, although reduced in area, is still present (cf. Herron, Dewey & Pitman 1974). If one reconstructs the Arctic region without any consideration to surrounding oceanic regions further south, it is possible to eliminate much of the area of the 'Palae-Arctic Ocean'. However, such an exercise produces insuperable problems in the reconstruction of the Atlantic.

The crust of the 'Palae-Arctic Ocean' would have been generated before the development of the present crust of the Arctic Ocean and, indeed, before the break-up of Pangaea. It would have to have been subducted during the Mesozoic and Tertiary but, as the author pointed out (Owen 1976), the geological evidence from the circum-Arctic Ocean region positively precludes such subduction during the last 200 million years. Map 6, the reconstruction at Anomaly 24 (Palaeocene), underlines the problem even further. In this map, substantial gores are shown to be still present only 56 million years ago. There is certainly no evidence to support the presence of major compressional subduction zones in the Arctic region during the Cenozoic. On the contrary, the evidence is of oceanic crustal growth originating initially at tensional margins. Maps 6, 9 and 12 here, together with those of Smith, Hurley & Briden (1981), which assume a constant modern dimensions Earth, while possessing cartographic integrity, do not comply with the geological and ocean-floor spreading evidence obtained from this region. The idea of Vink (1982), that these gores in the Boreal region and elsewhere can be explained away by an exceedingly fortuitous extension of the continental crust during displacement, is not supported by the field evidence.

If one subtracts oceanic crust in stages from the modern Boreal region (Map 4), while reducing the Earth's diameter systematically to a value of 10202 km (80% of its modern mean value) at the time of Pangaea (180–200 Ma), a series of reconstructions is produced which conforms to the field evidence (Maps 4, 5, 7, 8, 10, 11 and 13). This series invokes major displacement and tectonic interplay of the Alaskan and Anadyr–Verkhoyansk Blocks along known wrench faults and orogenic belts (Owen 1973, 1976; Jeletzky 1980).

The reconstructions at Anomaly 9 (Oligocene) and

Anomaly 24 (Palaeocene), shown in Maps 5 and 7 respectively, are fully controlled by the ocean-floor spreading data. This period of time covers the completion of spreading in the Amerasia Basin (Eocene) and the bulk of the development of the Eurasia Basin. Although there are regional constraints upon the possible geometric development of the Amerasia Basin, the precise stages of separation of the Canadian and Siberian margins are conjectural at present. The sequence of this separation shown in Map 8 (Turonian), Map 10 (Hauterivian) and Map 11 (Oxfordian) may need to be revised when sufficient spreading data become available. The development of the Canada Basin is due to the movement of Alaska in respect of both North East Asia and Canada. It is possible that this basin may have opened earlier than the main spreading phase that generated the crust of the Amerasia Basin, but it would have been subsequent to the middle Jurassic. The continental tectonic motions indicated are grossly over-simplified, but the trends shown are valid. Map 13 shows the reconstruction of Pangaea assuming an Earth with a diameter 80% of its modern mean value. Apart from the necessity to re-adjust the arrangement of the Canadian Arctic Islands, there are no gores present in the reconstruction and, by reversing the sequence of maps assuming an expanding Earth, no areas of anomalous crust appear or are required to be subducted.

The Palae-Arctic Ocean of reconstructions which assume a constant modern dimensions Earth at the time of Pangaea, even if it is partly eliminated by judicious movements of North East Asia, is a spherical geometric consequence of an Earth with a modern surface curvature. If one increases the value of surface curvature (the effect of reducing the diameter of the Earth), this gore is eliminated.

SECTION 2

North Atlantic

MAPS 14–23

The margins of the North Atlantic, like those of the Arctic, are tensional (passive) and preserve the full history of the initial continental splitting and the subsequent commencement of oceanic crustal generation. With the exception of the Puerto Rico Trench, no oceanic subduction zones are present and the spreading patterns provide, therefore, a relatively complete history of ocean basin formation.

The North Atlantic magnetic anomaly patterns have been mapped and interpreted in detail (see 'Data sources, handling methods and limits' above for sources) and the data are plotted diagrammatically on Map 14. The field data indicate that ocean-floor spreading commenced to separate the north-west African margin and the East Coast embayment of North America in the late middle Jurassic (Callovian), following upon an earlier Jurassic 'Red Sea' phase and rift-faulting extending back into the Triassic. As this southern region of the North Atlantic continued to widen during the upper Jurassic, rotating

North America and Africa away from each other, tension and continental splitting occurred further north. Differential motions between the northern region of Africa north of the Southern Atlas fault and southern Europe, produced major wrench motions and fold belts in the 'Mediterranean' region. These motions led, in the Kimmeridgian, to the start of an anticlockwise rotation of the Iberian Peninsula, which produced the Bay of Biscay oceanic crust by the early Tertiary.

During the late Jurassic and lower Cretaceous, the spreading axis extended northward between the Grand Banks continental shelf and Iberia to form a triple junction (a tensional point) close to the origin of the Gibbs fracture zone. This tensional regime, which continued in response to the progressive widening of the more southerly regions of the North Atlantic, extended limbs northward in two directions. The first limb was directed to the north-east to produce block faulting and rifting across southern England and northern France into the North Sea close to the Jurassic–Cretaceous boundary. This tensional line ran northward through the North Sea and affected the common margin of Greenland and Norway which then existed, re-activating older rift faults (e.g. Allen 1976; Haller 1969; Owen 1971, 1973; Surlyk 1978a, b).

The north-west limb extended northward and tension was sufficient to commence the development of ocean-floor spreading in the mid-Cretaceous between Bligh Bank, Rockall Bank and the continental margin of the British Isles (see Roberts, Masson & Miles 1981, Appendix Note 3). It also started to generate the crust of the Labrador Sea, situated between the Canadian and Greenland continental margins, in the very late Cretaceous. During the early Tertiary, oceanic crustal generation continued from the mid-Labrador Sea ridge (Ran ridge). This spreading axis penetrated further northward to form the Davis Strait and Baffin Bay and its tensional effects are also to be seen in the Amerasia Basin of the Arctic Ocean (see Section 1). Crustal generation ceased along this axis in the early Oligocene (Anomaly 13). The Davis Strait has an oceanic crust and is homologous with the 'hot spot' development of the Faeroes–Iceland–Greenland ridge.

In the early Palaeocene, a spreading axis developed along the former Mesozoic rift valley system which marked the Greenland–Scandinavian common continental margin. After the generation of the dykes of Anomaly 19 (Eocene), this axis became the sole generating ridge in the northern North Atlantic. The ridge, known successively as the Reykjanes–Iceland ridge and the Mohns–Knipovich ridge, has generated the Denmark Strait and the Norwegian–Greenland Sea. Although offset by the Spitzbergen fracture zone, this spreading axis penetrated into the Arctic where it is known as the Nansen (or Gakel) ridge, from which has been generated the Eurasia Basin of the Arctic Ocean (see Section 1).

The foregoing account is, in outline, the spreading history of the North Atlantic which can be deduced from the data plotted on Map 14. The North Atlantic can be visualized as a spherical triangle, the early base of which is marked by the Bahama–Doldrums fracture zone. As the base widened, so the apex of the triangle extended northward separating North America from Africa, initially, and from the European–Greenland margin subsequently. Continued widening to the south caused the apex to extend firstly between Canada and Greenland and then between Greenland and Europe, into the Arctic. It is now necessary to examine the reconstructions to establish whether the constant modern dimensions Earth hypothesis or the expanding Earth hypothesis, or indeed both, will satisfy the field data.

The reconstructions of the North Atlantic which assume a constant modern dimensions Earth are shown in Maps 16, 19 and 22. Removal of all oceanic crust generated after Anomaly 24 (Palaeocene) and re-assembly of the remainder on a scale modern-size globe, produces the arrangement shown in Map 16. The Anomaly 24 reconstruction is weighted to give the best fit within the North and South Atlantic. Nevertheless, an unavoidable narrow gore has to be constructed in the region of the Azores–Gibraltar fracture zone. But it is in the Labrador Sea region and further north (Map 6) that gores are readily apparent. Major problems of fit occur in the Anomaly M7 (Hauterivian) reconstruction given in Map 19, apart from the gore in the Arctic region discussed previously (Section 1) and shown on Map 9. If one assumes a modern dimensions Earth, it is necessary to overlap the spreading patterns in the southern North Atlantic in order to eliminate the gores in the region north of the Newfoundland–Gibraltar fracture zone. It can be argued that this problem can be solved if Africa and South America are rotated together clockwise, sufficiently to eliminate the gore in the South Caribbean and remove the overlap in the spreading pattern in the southern North Atlantic. Unfortunately, this would widen the Tethyan Ocean to an extent greater than at the time of Pangaea and at a time when that Ocean should be contracting in area by subduction of its crust. It would also extend the apex of the Tethyan Ocean westward as a spreading zone into the western Mediterranean region.

The classic fit together of the continents bordering the subsequent North Atlantic is shown in the reconstruction of Pangaea (Map 22). It is a reasonably precise fit at the centre of the re-assembly where the north-west African margin fits into the North American East Coast embayment. However, gores are present between Greenland and the Canadian Arctic Islands in regions such as Baffin Bay and the Nares Strait. In the latter case, the geological evidence is of direct contact between the Ellesmere Island and Greenland continental margins at a wrench fault trace, without any evidence of the necessary Mesozoic or Cenozoic crustal subduction (see also Map 12). The gore which widens southward between Africa and South America is discussed below (Section 3).

It is readily apparent that the ocean-floor spreading data do not coincide with the spherical geometric requirements of reconstructions which assume an Earth with a surface curvature of that of the modern Earth and thus of the same diameter. The reconstructions which assume an expanding Earth show a development of the North Atlantic region which, with the changing value of surface curvature, allows of the fit together of isochronous pairs of magnetic anomalies and a fit of the continents themselves at the time of Pangaea in accordance with the geological evidence (Maps 14, 15, 17, 18, 20, 21 and 23).

SECTION 3
South Atlantic

MAPS 24–33

The continental–oceanic margins of the South Atlantic are tensional (passive), like those of the North Atlantic. A full history of continental splitting and ocean basin development is discernible. One small subduction zone is present, marked by the South Sandwich Trench, which reflects the interaction of a growing Scotia Sea oceanic crust with continued spreading of the South Atlantic.

The magnetic anomaly patterns that have been mapped and identified are shown plotted on Map 24 (see 'Data sources, handling methods and limits' above for sources). If one subtracts progressively the isochronous magnetic anomalies on a globe until the fit of South America and Africa is achieved, a good picture of the development of the South Atlantic is obtained. Although now a continuous spreading zone with the North Atlantic, the early history of the South Atlantic shows that these two oceans were quite separate until the mid-Cretaceous. Unfortunately, there is one relatively small but important area, the Weddell Sea, for which there was no spreading information available at the time the reconstructions were mapped. The development of the area until the late Cretaceous shown here is conjectural, but is consistent with the relative motion of West Antarctica to Africa and South America in response to the development of the South Atlantic and Indian Ocean.

Spreading between southern Africa and South America commenced in the early Cretaceous (Valanginian), following the line of an earlier, Jurassic, phase of rift-faulting. Like the North Atlantic, the base of the developing spherical triangle of oceanic crustal generation was situated in the south and is marked by the now separated Agulhas and Falkland fracture zones. The apex of the triangle extended rapidly northward during the lower Cretaceous and, by the late lower Cretaceous, a triple junction had formed close to the Nigerian/north-east Brazilian continental margin. The north-east spreading limb extending from this junction, formed the oceanic floor of the Benue Trough overlain by Albian sediments. The north-west limb extended between the Guinea Coast and the North Brazilian continental margins. The subsequent history of the west–east spreading region of the South Atlantic, is one of progressive widening with a clockwise rotation of South America in response to the greater area of oceanic crust generated in the south rather than in the north. From the Turonian onward, the crustal generation of the South and North Atlantic occurred from a common mid-Atlantic ridge.

The production of the oceanic crust of the South Atlantic, south of the Falkland–Agulhas fracture zone, is intimately linked with the development of the western part of the Indian Ocean. The displacement of Madagascar southward and away from Africa to form the Somali Basin and Mozambique Channel commenced in the middle Jurassic and was completed by the late lower Cretaceous (see Section 4). This displacement, while being partly in response to the anticlockwise rotation of Africa due to spreading in the North Atlantic, is mainly due to a relatively clockwise rotation of Antarctica and Australia together, in respect of Africa. By the end of the lower Cretaceous, the regions of the Falkland Plateau and the corresponding Agulhas Plateau (south of Africa) were being separated by north–south trending anomalies, half the pattern of which has been mapped in the Georgia Basin. The southern margin of this separating strip of oceanic crust is marked by the Scotia fracture zone and by a now-dislocated continuation of the Mozambique fracture zone.

South of this joint fracture zone, the Weddell Sea would have developed in response to the stretching and eventual separation of the West Antarctic Peninsula from East Antarctica, in response to the early development of the South Atlantic and western Indian Ocean basins (see Kellogg 1980, Appendix Note 4). The clockwise rotation of East Antarctica relative to Africa, produced the Indian–Antarctic basin spreading zone with its lower Cretaceous anomaly sequence, one half of which can be seen off Droning Maud Land and the other half in the Mozambique Basin.

The interplay of the effects produced by spreading in both the South Atlantic north of the now-disrupted Falkland–Agulhas fracture zone and in the Indian Ocean, is well seen in the late Cretaceous spreading pattern mapped in the Agulhas Basin. Continued clockwise rotation of Antarctica away from Africa and India led eventually to the transformation of the former Scotia fracture zone to a spreading axis, which has generated the crust of the Scotia Sea during the Cenozoic and is still active today. The spreading pattern marking the north–south separation of Africa and Antarctica indicates that its generation was episodic during the Cenozoic. Even now, the South West Indian ridge spreading axis has barely broken through the older spreading patterns.

The development of the North and South Caribbean is intimately connected with the motions of South America and Africa relative to North America and it is appropriate to consider this here. North America has been displaced away from Africa from the Callovian (late middle Jurassic) onward. However, South America and Africa were displaced together relative to North America during the upper Jurassic and lower Cretaceous and it was not until the Albian (late lower Cretaceous), some 50 million years later, that these two continents became fully separated. Despite the clockwise rotation of South America shown by the spreading patterns in the South Atlantic and Pacific, the sum effect of the continental motions has been to elongate the central American and Caribbean region in a north-west to south-east direction. The North Caribbean developed first in the upper Jurassic and lower Cretaceous with some subduction at the Greater Antilles line. The South Caribbean developed in the upper Cretaceous and lower Tertiary with Cenozoic motions of South America producing subduction at the Venezuelan margin, the Lesser Antilles ridge and, more recently, at the Puerto Rico Trench. The displacement of the Nicaragua–Honduras craton relative to the western margin of Mexico and Yucutan advocated previously (Owen 1976) and here, is supported by the geological trends and tectonic patterns in this region.

This outline account of the spreading history of the South Atlantic conforms to the magnetic anomaly patterns present. It now remains to examine the reconstructions which assume a constant modern dimensions Earth and those which assume an expanding Earth, in order to see whether they agree with this reading of the field data.

The reconstructions which assume a constant modern dimensions Earth, are deliberately weighted to give the best possible fit in the South Atlantic region (Maps 26, 29 and 32). The fit at Anomaly 24 (Palaeocene), shown in Map 26, agrees closely with the spreading data except in the region of the St Paul's, Romanche, Chain and Fernando de Noronha fracture zones. The South Caribbean region is rather too narrow in the north–south direction. But, this fit at Anomaly 24 is an example of concentrating on providing the best fit in the region under consideration and ignoring problems of fit elsewhere. If one turns to the corresponding map of the Indian Ocean (Map 36), major gores are apparent. The partial removal of these anomalous gaps in the Indian Ocean would be at the expense of the fit in the South Atlantic.

The lower Cretaceous Anomaly M7 (Hauterivian) reconstruction shown in Map 29, is also in agreement with the field data in the region of the South Atlantic. However, a gore is shown in the region of the South Caribbean and overlap of oceanic crust occurs in the southern North Atlantic. These misfits can be eliminated if a clockwise rotation is made of Africa, the early Indian Ocean and Antarctica. But, this can be achieved only if the Tethyan Ocean is widened significantly to a greater area than at the time of Pangaea, with all the attendant problems of subsequent subduction rates at the Alpine–Himalaya belt (for a more complete statement of the problems see Section 4).

In the case of the reconstruction of Pangaea (Map 32), a distinct gore is unavoidable widening southward between South America and Africa, shown by dark shading. This can be eliminated by rotating South America in an anticlockwise direction. However, doing so merely transfers the gap to the region between the Guinea and North Brazilian margins. In the case of either gore, the crust would be older than late Triassic and would have to be subducted at some time before the onset of the oceanic crust actually present today. There is no evidence of Jurassic subduction zones at the passive continental margins concerned.

If one now turns to examine the reconstructions which assume an expanding Earth (Maps 24, 25, 27, 28, 30, 31 and 33), the development of the South Atlantic shown by them agrees with the spreading data. The development also agrees with the ocean-floor spreading patterns in the North Atlantic and in the Indian Ocean without any problems of gores or overlaps. In accord also, is the development of the Caribbean region.

SECTION 4

Indian Ocean

MAPS 34–43

The Indian Ocean shows three distinct phases of ocean-floor spreading (Map 34). The first phase extended definitely from the Oxfordian (upper Jurassic) and possibly from the Callovian (middle Jurassic), up to the middle of the upper Cretaceous. It produced the bulk of the Somali Basin (which might have an earlier origin), the Mozambique Channel and the Indian–Antarctic Basin, lying to the west of India, and the north-eastern and southern regions of the Wharton Basin to the north and west of Australia. This initial spreading phase reflects the relatively clockwise motion which rotated the combined Antarctic–Australian continent away from Africa, Greater India and South East Asia. To this is added the effect of the anticlockwise rotation of Africa in response to spreading in the early southern North Atlantic. The combined effect of these rotational movements in the western region is shown by the development of the Somali Basin and the associated displacement of the Island of Madagascar relative to Africa along the trace of the Davie fracture zone. Madagascar also moved southward relative to Greater India, which in turn has been displaced southward relative to the southern margin of Asia, but to a far lesser extent.

The pivot of the rotational movement of the combined Antarctic–Australian continent appears to have been situated close to the Transantarctic Mountains, separating East from West Antarctica, along which there has been some wrench motion. The movement of Australia away from South East Asia from the early upper Jurassic onward and from Greater India in the lower Cretaceous, produced the pattern of spreading seen to the west and north of the tensional Australian continental margin. The other half of this spreading pattern has now been subducted at the Mentawai–Java Trench system, which forms the only subduction zone in the Indian Ocean, the remaining margins being passive with a full splitting and spreading history preserved. The delay in the separation of Australia from Greater India is particularly interesting, as there is evidence of some ocean-crustal generation in the late Jurassic and lower Cretaceous between the northern margin of Greater India and the southern margin of Asia (cf. Stöklin 1977, 1980).

The second phase of ocean-floor spreading in the Indian Ocean region extended in time from the mid-upper Cretaceous to the Palaeocene. The patterns evident in Map 34 indicate that generation of oceanic crust occurred at a west–east trending spreading axis. This spreading axis was offset at fracture zones and, particularly, by the Ninetyeast ridge transform fault. Although a clockwise rotation of the combined Antarctic–Australian continent is still apparent in the pattern, the principal feature of this second phase of ocean-crustal generation is the north–south widening of the Indian Ocean. With this widening, there is associated a northward displacement of Greater India and the probable stretching southward of the South East Asian, Philippines and Indonesian regions.

During the Palaeocene, a southern Pacific spreading axis extended westward between Australia and Antarctica, beginning the separation of these two continents and the third spreading phase evident in the Indian Ocean. As the post-Palaeocene crust grew between Antarctica and Australia, the latter continent was displaced northward towards South East Asia, Indonesia and the flanking Pacific region. The original northern region of the Wharton Basin oceanic crust was eliminated by the development of the Mentawai–Java Trench subduction system. Widening of this third spreading region in the east, led to the rapid penetration of the apex of the spherical triangle westward to cut across the older spreading patterns. Initially, in the Eocene, the axis reached as far west as the former transform fault trace, the Chagos–Maldive–Laccadive fault system, which had marked the western limit of the northward movement of India during the Cretaceous and early Tertiary. Tension along this system caused it to become a 'leaky' transform with the building up of a ridge system by volcanic activity. Tension also caused the former Ninetyeast fracture zone, which in turn marks the eastern trace of the northward displaced Indian 'plate', to become 'leaky' and to build up into a wide ridge system.

Tension produced two triple junctions in the Miocene. The first of these was situated at the southern end of the Chagos ridge, one limb sending the spreading axis northward and north-westward along the line of the Central Indian Ocean ridge and the Carlsberg ridge, to reach and form the Gulf of Aden. The other limb is represented by the spreading axis which split across the older spreading patterns towards the South Atlantic, to form the South West Indian ridge. The second triple junction was also formed in the Miocene at the western end of the Gulf of Aden spreading axis, the other two limbs being represented by the actively spreading Red Sea axis and the East African Rift Valley system.

The description of the development of the Indian Ocean given above is readily apparent from the information plotted on Map 34. To what extent do the constant dimensions Earth reconstructions coincide with the spreading data? In the discussion of the corresponding South Atlantic maps (see Section 3), it was pointed out that the fit together of the pair of Anomaly 24 dykes was at the expense of the fit in the Indian Ocean. Map 36 shows a major gore which has to be constructed in the region of the Ninetyeast ridge. This gore widens northward to join up with a substantial eastern area of Tethyan Ocean crust which has still to be subducted and, of which, only the lighter grey shaded areas can be justified. A large area of Tethyan Ocean crust lying to the north of India has also to be subducted. Within the Indian Ocean itself, rather less obvious regions such as the Chagos–Maldive–Laccadive ridge system and Kerguelen have to be included in their later configuration and are, therefore, anomalous. The amount of subduction required to remove the remaining area of Tethyan Ocean crust between Anomaly 24 and the main Oligocene phase of the Alpine–Himalaya orogen makes this reconstruction untenable (see also 'Some errors in reconstructions' above and Appendix Note 5).

The Anomaly M7 (Hauterivian) reconstruction of the

Indian Ocean region (Maps 29 and 39) is consistent with the spreading data. However, Antarctica is some 25° of latitude further north than it should be according to the data fixing the position of the south magnetic pole, although this can be improved by 12° of latitude if the reconstruction is oriented so that the north geographic pole of the fixed co-ordinate graticule is made to coincide with the north magnetic pole indicated on Map 9. The remaining latitudinal discrepancy cannot be made good by rotating South America, Africa, Antarctica and Australia, with the intervening oceanic regions, clockwise through an arc of approximately 12° in order to eliminate the overlap in the southern North Atlantic seen in Map 19. This latter rotation would produce a much wider Tethyan Ocean than at the time of Pangaea, but, the closure of this ocean by subduction throughout its length during the upper Cretaceous and early Tertiary poses major problems concerning the reconciliation of the ocean-floor spreading patterns and the subduction rates required. There is some evidence of crustal tension and even of ocean-floor spreading at the margins of Greater India and southern Asia during the late Jurassic and the Cretaceous up to about the Turonian, but this is more in keeping with the expanding Earth model discussed below.

The reconstruction of Pangaea on a constant modern dimensions Earth requires the presence of a substantial triangular-shaped crustal area called the Tethyan Ocean. There is no other spherical geometric course but to construct this gap separating Arabia, India and Australia from southern Asia and South East Asia (Map 42). In 1976, the author pointed out that this 'ocean' was a necessary geometric artefact and that geologists who had worked extensively in the Alpine fold belts from the near-east to Burma, had denied the presence of any extensive oceanic area in the late Triassic and early Jurassic (e.g. Spencer 1974 and, more recently, Stöklin 1977, 1980). However, the shortening of crustal area along this fold belt is evident enough during the later Mesozoic and in the Cenozoic.

The reconstructions which assume an expanding Earth, show a logical kinematic sequence of development of the Indian Ocean (Maps 34, 35, 37, 38, 40, 41 and 43), which is in accord with the spreading data shown on Map 34 and discussed above. The re-assembly of the continents to form Pangaea, assuming an Earth 80% of its modern mean diameter, eliminates the Tethyan Ocean gore (Map 43). This does not mean that a Tethys Sea did not exist at this time. On the contrary, the evidence is of an extensive epicontinental sea present in the Triassic.

The development of the southern half of the spreading patterns in the Wharton Basin off Australia during the upper Jurassic and lower Cretaceous, is well documented (Maps 41 and 40). Assuming that spreading was more or less symmetrical in this region, one can postulate the area and configuration of the other half of the spreading pattern now subducted and this is shown as a lightly shaded area on these maps. The displacement of Africa during the upper Jurassic and lower Cretaceous, in accordance with the spreading patterns in the Atlantic, requires the production of a gap between Arabia and Turkey. Greater India, also, is stretched in the north–south direction as it tends, initially, to rotate together with

Antarctica, thereby assisting in the development of the Somali Basin and the Mozambique Channel. This post-middle Jurassic break-up of the Indian Ocean region of Pangaea involves, therefore, significant tension between the northern margin of Greater India and the southern margin of Asia. This tensional belt may have been a zone of ocean-floor spreading and would account for the presence of the ophiolite zones and very deep water sediments of late Jurassic and lower Cretaceous age to be found deformed in the Alpine–Himalaya chains stretching from Turkey to Burma (e.g. Stöklin 1977). The axis of this spreading zone would be related, on the one hand, to that which had generated the oceanic crust in the Eastern Mediterranean and, on the other, to that which had generated the spreading patterns to the north and north-east of the Exmouth Plateau in the Wharton Basin (and compare the trend of the Phoenix Plate of the Pacific shown in Maps 50 and 51). From the Turonian onward (Maps 38, 37, 35 and 34), the earlier tensional phase is replaced by a compressive and perhaps subductive phase, as Greater India was displaced relatively northward.

The remaining sequences of development, shown in Maps 37, 35 and 34, coincide precisely with the spreading data and the subduction history at the Mentawai–Java Trench system.

SECTION 5

Pacific

MAPS 44–53 & 54–63 (part)

The central and northern regions of the Pacific are shown graphically in Maps 44–53 and the southern region in Maps 54–63, an overlap between 20° and 30° south latitudes being provided in each corresponding map. Unlike the oceans previously considered, the continental margins of the Pacific are, or have been, tectonically active in a compressional and thrusting role. The margins are marked by wrench fault traces and major subduction zones. The determination of a complete and precise history of ocean basin development is not possible, therefore, in the absence of a substantial region of oceanic crust, the generation of which can be inferred with confidence.

The modern maps (Maps 44 and 54) show that a considerable amount of magnetic anomaly information is available, with a detailed coverage in some areas. They demonstrate, also, that if spreading has been symmetrical, nearly half of the oceanic crust of the Pacific generated from the middle Jurassic onward, has been subducted at, or near, the ocean margins. There is strong spherical geometric evidence which indicates the former presence of a very large area of pre-Triassic oceanic crust, now totally subducted. In order to understand the development and destruction of the Pacific Ocean crust during the Mesozoic and Cenozoic, it is necessary to consider the effect on this region of the break-up of Pangaea and the subsequent displacement of the continental fragments by the formation of the Arctic, Atlantic and Indian Oceans.

Whether one assumes a constant modern dimensions Earth, or an expanding Earth, the grouping of the continents together in Pangaea produces a substantial area of Pacific Ocean crust at the time of the late Triassic to early Jurassic (e.g. Maps 52 and 53, 62 and 63). The late middle Jurassic commencement of ocean-floor spreading in the southern North Atlantic started the displacement of North America westward, overriding in the process the older Pacific oceanic crust. This process has continued throughout the remainder of the Mesozoic and the Cenozoic up to the present day. The displacement westward of South America at the expense of the Pacific did not commence until some 50 million years later than that of North America, within the early lower Cretaceous. This time delay in westward displacement is reflected in the modern Pacific by the presence of a substantial remnant of the post-Cretaceous oceanic crust generated to the east of the main Pacific spreading axis. Apart from the westward displacement of the Americas, there is a relatively minor displacement of the continental regions of the Arctic into the former North Pacific area, and a major incursion produced by the development of the Indian Ocean crust since the late middle Jurassic. These displacements into the region of the 'Eo–Pacific', together with the post-middle Jurassic generation of oceanic crust in the Pacific region itself, have caused the elimination of all pre-middle Jurassic crust.

The spreading patterns in the modern Pacific indicate that the crust which remains today, commenced to form in the lower or middle Jurassic from a triple junction situated to the south-west of the Mid-Pacific Mountains. The triple junction, itself, will now have been subducted but, from the palaeomagnetic latitude data from the lavas of the subsequent spreading pattern, it can be shown by extrapolation to have been situated in the South Pacific around 30–40° South latitude. By Anomaly 23 (Oxfordian), the initial spreading limbs extending from the triple junction had expanded the oceanic crust to initiate the Japanese, Hawaiian and Phoenix plates. The spreading axis of the Japanese plate trended west–south-west to east–north-east and linked with the north–north-west to south–south-east trending spreading axis of the Hawaiian plate at a distinct magnetic bend analogous to the Great Magnetic Bight seen in the later Cretaceous and early Tertiary pattern in the north-east Pacific. Only the southern half of the Japanese plate spreading region is preserved, but it is interesting to note that the Mesozoic pattern in the Bering Sea has a not too dissimilar trend and represents the north-western half of a former spreading zone.

The Phoenix plate is also the northern half of a spreading zone, the southern portion of which has been subducted, probably at the northward advancing Australasian margin. It possesses a similar trend to the contemporary spreading pattern in the Wharton Basin of the Indian Ocean to the north and east of the Exmouth Plateau (Section 4, Maps 38, 40 and 41). Palaeomagnetic latitude determinations indicate that in the lower Cretaceous (about 120 Ma), the Phoenix plate was some 40° of modern latitude south of its current position. Larson & Chase (1972) considered that the basic relationship of the Phoenix, Hawaiian and Japanese plates had not altered since their crustal generation. The northward displacement of this Jurassic and lower Cretaceous spreading zone as a whole, appears to be certain.

The middle and early upper Cretaceous oceanic crust flanks the Hawaiian and Phoenix plates, indicating that spreading continued from the same generating axes. In the case of the Japanese plate, most of the post-lower Cretaceous oceanic crust has been subducted at the Kuril–Kamchatka Trench. A comparison of the spreading patterns west and east of the Emperor Seamounts suggests, strongly, that this chain together with the Hess ridge and, possibly, the Hawaiian ridge is a major transform fault. This line demarcates a westerly crustal region which has been displaced northward in response to the post-Palaeocene migration of Australasia away from Antarctica. Such a displacement indicates that the Great Magnetic Bight is strictly homologous to the similar, but much earlier (Jurassic) bend between the Japanese and Hawaiian plates (see Farrar & Dixon 1981, Appendix Note 6).

If one accepts this interpretation, it follows that the west–east spreading axis which was situated to the south of the Aleutian Trench during the late upper Cretaceous (Anomaly 32B onward), was an extended version of the same axis which had generated the Japanese plate. The major north–south trending spreading axis which generated the bulk of the oceanic crust of the eastern, north and central Pacific during the late Cretaceous and Cenozoic, is a greatly extended version of the same axis which generated the Hawaiian plate. The northern portions of this spreading axis has now been overridden by the western margin of northern North America.

Further to the south in the eastern Pacific, this spreading axis extended during the early upper Cretaceous to commence the separation of New Zealand from Antarctica. Part of the eastern half of this spreading zone is still preserved. This axis in turn, albeit offset by the Macquarie ridge transform fault, was responsible for the late Cretaceous (Anomaly 32) to Palaeocene (Anomaly 24) generation of the Tasman Sea which separated New Zealand and the Campbell Plateau from Australia. From the Palaeocene onward, this axis changed direction to penetrate between Australia and Antarctica and commence the generation of the third phase of spreading which formed the crust of the Indian Ocean (see Section 4). During the Miocene, a new spreading axis partially split through this older late Cretaceous and Cenozoic spreading zone in the South Pacific, to generate a spherical triangle of oceanic crust from the western apex of the triangular Galapagos spreading zone southward. This spreading axis continues to generate oceanic crust at the present day. Tension in the south-east corner of the Pacific, associated with the same processes which led to the development of Drake Strait and the Scotia Sea spreading region, produced a triple junction. The most obvious Pacific spreading limb extends northward towards the Galapagos spreading zone and much of the axis is now extinct. However, there is an earlier Cenozoic spreading limb which has been partially overridden by an advancing West Antarctic margin (see Cande Herron & Hall 1982, Appendix Note 7).

The remaining late Cenozoic spreading region to consider in outline is that of the Galapagos. It is not readily apparent why this spreading zone should have developed, it being a tensional feature in a compressional region. If one assumes an expanding Earth model, however, a relatively westward motion of South America to the Caribbean region is required between Anomaly 9 (e.g. Map 15) and the present day (e.g. Map 14) in response to spreading in the northern South Atlantic. This movement would buckle the Panama Isthmuth and produce a tensional line in the Pacific which could lead, subsequently, to the formation of a spreading zone.

The development of the western Pacific marginal basins has taxed the ingenuity of a number of plate tectonicists to explain their essentially tensional nature in a generally compressional crustal regime. These marginal basins comprise the Sea of Okhotsk, the Sea of Japan and the South China Sea, off continental Asia; the Coral Sea and Fiji Basin (see Malahoff, Feden & Fleming 1982, Appendix Note 8), together with the smaller Cenozoic basins of Australasia; and one may include also, the Philippine and Caroline Basins of the Pacific itself. It is difficult to understand why these regions should have developed if the Earth has been of constant modern dimensions throughout the late Cretaceous and Cenozoic. In this model, the Pacific Ocean is subject to a continuous reduction of area in response to the development of the Atlantic, Arctic and Indian Oceans. There is little room for the necessary tensional regimes, at least to initiate the onset of ocean-floor spreading in the western Pacific.

If one turns to the constant modern dimensions Earth reconstructions (Maps 44, 46, 49, 52, 54, 56, 59 and 62) and those which assume an expanding Earth (Maps 44, 45, 47, 48, 50, 51, 53, 54, 55, 57, 58, 60, 61 and 63), there can be no rigorous control on the accuracy of either series of reconstructions because of the subduction at the Pacific margins. Up to the late upper Cretaceous, prior to Anomaly 34, the exact position of the remnants of the Jurassic and earlier Cretaceous spreading patterns is uncertain. From Anomaly 34 onward, there is some control on the possible position of the developing spreading patterns, because of the mode of separation of New Zealand from Antarctica and Australia.

The question is, which of these two series most closely satisfies the field data? Known gores in the Arctic and Indian Ocean regions in constant modern dimensions reconstructions, extend and widen towards the Pacific. In the series provided by Maps 44, 46, 49 and 52 and Maps 54, 56, 59 and 62, the area which is required to be subducted, assuming an expanding Earth, is shown in light shading. The remaining crustal area to be subducted in these constant dimensions reconstructions, is shown unshaded. The rates of subduction in the Pacific prior to Anomaly 24 (Palaeocene) are not known and so, the subduction rate required in these constant modern dimensions Earth reconstructions could be possible. Neither is it readily apparent why the older Jurassic and lower Cretaceous spreading region should have developed and been displaced northward in the manner indicated by the palaeomagnetic latitude data. There are problems, however, in getting rid of excess oceanic crust between Anomaly 24 (Maps 46 and 56) and the present day (Maps 44 and 54), which cannot be explained away by invoking a faster subduction rate during this period of time. To add to this, one also has to take into account the development of the western Pacific marginal basins, mentioned above, which is difficult to explain in the context of a steadily contracting Pacific region.

If one assumes an expanding Earth with a diameter 80% of its modern mean value at the time of Pangaea, it is necessary to construct a substantial area of oceanic crust which can be called the Eo-Pacific Ocean (Maps 53 and 63). This oceanic crust would have been generated during the Palaeozoic. The triple junction which initiated the generation of the Pacific Ocean crust seen today was sited at about 35° south latitude. The latitudinal position can be determined by extrapolation from the palaeomagnetic data obtained from the subsequent Phoenix, Hawaiian and Japanese plates, although the site of the triple junction has been subducted. It so happens that if one retains the relative position of these three growing plates to Australia in the reconstructions, their northward growth and subsequent displacement to their modern positions is in full accord with the palaeomagnetic field data. The trend of the spreading axes of these three plates is consistent with a Pacific Ocean growing in area as well as the mere increment of continuously generated oceanic crust (Maps 53 and 63, 51 and 61, 50 and 60, 48 and 58). Despite the global increase in surface area of an expanding Earth, North America and, eventually, South America would still be displaced by the growing crust of the Atlantic against the spreading trend of the growing Pacific. The oft quoted comment that all exponents of an expanding Earth deny the presence of subduction zones, or that subduction of oceanic crust has occurred, is obviously not correct in the case for global expansion discussed here. The continental margins of the Americas would certainly be overriding the older crustal areas of the eastern Pacific during the Mesozoic, by inference, and during the Cenozoic, by observable fact.

A similar subduction effect is to be seen in the western Pacific, especially with the displacement of Australasia during the Mesozoic and Cenozoic in response to the development of the Indian Ocean. The relative displacement of Asia towards the Pacific was considerably less than that of the Americas, however, and this is reflected in the marked asymmetry in the preservation of the Mesozoic and Cenozoic spreading patterns, with only the western half relatively complete and barely three-quarters of the total pattern preserved. As J. Steiner has indicated (1977), the rate of growth of oceanic crust during the Mesozoic and Cenozoic is nearly exponential. The style and rates of subduction required in the reconstructions which assume an expanding Earth, are consistent with the spreading rates in the Pacific, the generation of oceanic crust in the other oceans, and the increase in surface area (Maps 53 and 63, 51 and 61, 50 and 60, 48 and 58, 47 and 57, 45 and 55, 44 and 54). The spreading and subduction history of the Pacific Ocean outlined at the beginning of this section, coincides with the reconstructions mapped here which assume an expanding Earth. Nonetheless, the detailed arrangement of spreading and subduction patterns is likely to be the subject of some future revision.

It has been mentioned above, that the developments of the western Pacific marginal basins are better explained by an expanding Earth model than by the conventional constant modern dimensions Earth model. These are now considered in more detail.

The Sea of Okhotsk was probably formed during the Cretaceous in response to the displacement of the Anadyr–Verkhoyansk Block of north-east Asia by the growing Amerasia Basin of the Arctic Ocean (Maps 8, 7, 5 and 4). The later stages of this displacement southward would de-couple the original marginal fold belt of east Asia from its cratonic hinterland, producing the Sea of Japan oceanic crust by tensional spreading. What is known of the trend of the magnetic anomalies in the Sea of Japan supports this assertion.

In the case of the South China Sea and the various oceanic crustal basins in the Philippines themselves, the cause of their development is the southward stretching of this entire region of South East Asia in response to the development of the Indian Ocean (Maps 43, 41, 40, 38 and 37). The subsequent buckling of the eastern region of the Philippines due to the northward displacement of Australasia after Anomaly 24 (Maps 35 and 34) has also produced tensional basins during the Cenozoic.

The markedly greater increase in oceanic crustal growth in the southern hemisphere during the late Cretaceous and Cenozoic, apparent in Map 54, provides the clue to the reason for the southern Pacific spreading axis extending to form the Tasman Sea between Australia and New Zealand. The author has stated (Owen 1976), that the ocean-floor spreading evidence is consistent with an Earth which has bellied out southward during the upper Cretaceous and Cenozoic, while retaining its shape as a rotational sphere. This reading of the spreading patterns world-wide, is supported by the apparent northward migration of the oceanic crust during the upper Cretaceous and Cenozoic according to the palaeomagnetic latitude data. An Earth expanding in this manner will have a magnetic field equator and a climatological equator which will move southward, progressively in relation to previously formed oceanic crust, during the later Mesozoic and Cenozoic, to arrive at its present position. In terms of the magnetic inclination preserved in the lavas generated at a given place and time, those intruded in the Phoenix plate in the early Cretaceous, for example, show that the sites were then in the southern hemisphere, while now they are situated in the northern hemisphere (compare Maps 50 and 60 with Maps 44 and 54). However, the Earth would appear to enjoy confusing the ocean-floor spreader. On top of this apparent northward migration, there has to be added a substantial tectonic motion of the crust of the Phoenix, Hawaiian and Japanese plates in the same direction, due to the post-Anomaly 24 (Palaeocene) displacement of Australasia into the western Pacific and away from Antarctica.

The increasing area of the South Pacific reflects a predominantly tensional regime. During the late Cretaceous and early Tertiary, the crust of the Tasman Sea and the Coral Sea was generated as Australia was still being displaced relatively southward. But, although from the Palaeocene onward, Australasia has been displaced

northward against the Pacific Ocean crust, tension is still apparent, producing the Cenozoic basins, such as the Fiji Basin, by ocean-floor spreading (see Malahoff, Feden & Fleming 1982, Appendix Note 8). The displacement of Australasia northward has in turn caused the Phoenix–Hawaiian–Japanese spreading region to be displaced northward, the northern margin of the Japanese plate being subducted at the Kuril–Kamchatka Trench. The eastern margin of this zone is marked by the Emperor Seamounts chain, Hess ridge and part of the Hawaiian ridge (see Farrar & Dixon 1981, Appendix Note 6). To the west of this Jurassic and lower Cretaceous spreading region, lies a series of basins, including the Caroline Basin and the Philippines Sea, of Cenozoic age. If one examines the three reconstructions mapped in Maps 47, 45 and 44, it becomes apparent that the northward displacement of the Phoenix, Hawaiian and Japanese plates would imply subduction in some areas, but tension in others. The magnetic anomaly patterns of these marginal basins are in full accord with these tensional directions.

SECTION 6

Southern oceans

MAPS 54–63

It is instructive to examine the spreading zones surrounding Antarctica from a position above the south geographic pole. A substantial amount of new data is now available since the author's previous review of the ocean-floor spreading patterns and their spherical geometric implications (Owen 1976). The spreading data are plotted on Map 54 and have already been discussed in the descriptions of the development of individual ocean basins. In essence, it is the displacement history of Antarctica away from South America, Africa and Australia with the development of the southern South Atlantic, Indian and Pacific Oceans, that is now considered.

The simple act of removing isochronous areas of oceanic crust in sequence from around Antarctica, leads eventually to a configuration of Gondwanaland. However, Antarctica could not have retained its position over the geographic south pole at the time of Pangaea, without the production of a wide gap between Gondwanaland and Laurasia and the disruption of the known fit of North West Africa into the North American East Coast embayment.

If one examines the series of reconstructions which assume an Earth of constant modern dimensions (Maps 54, 56, 59 and 62), a number of problems become apparent. At Anomaly 24 (Palaeocene), the reconstruction shown in Map 56 requires the presence of two substantial gores. One of these is at the position of the Ninetyeast ridge, the other to the north-west of New Guinea. If one attempted to close the gore at the Ninetyeast ridge, it would require an anticlockwise rotation of East Antarctica, Australasia and the Wharton Basin to do so. However, this rotation would, in turn, open up a gore of equivalent extent between West and East Antarctica. The presence of either gore is not supported by the ocean-floor spreading and geological evidence; the reconstruction is not tenable.

In the case of the Anomaly M7 (Hauterivian) reconstruction shown in Map 59, the palaeomagnetic evidence indicates that Antarctica was situated over the south magnetic pole. By comparison with the map of the Boreal region at Anomaly M7 (Map 9), which assumes a modern dimensions Earth, the position of the north magnetic pole is approximately 12° southward of the north geographic pole, constructed here, close to the 160° west meridian. If one assumes that the Earth's magnetic field was perfectly dipolar, the south magnetic pole would be at approximately 12° north of the south geographic pole on the 20° east meridian. Antarctica still fails to straddle this magnetic pole and neither would it if one rotated Gondwanaland and the intervening early Indian Ocean crust clockwise, to remove the crustal overlap in the southern North Atlantic (Map 19).

The series of reconstructions which assume an expanding Earth (Maps 54, 55, 57, 58, 60, 61 and 63), coincide with the ocean-floor spreading and palaeomagnetic data throughout the Mesozoic and Cenozoic.

The test of the ocean-floor spreading data given in this atlas indicates that, if one assumes an Earth of constant modern dimensions throughout the Mesozoic and Cenozoic, the reconstructions cannot be made to coincide fully with the spreading data. On the other hand, if one assumes that the Earth has increased its diameter from a value of 80% of its modern mean length 180–200 million years ago, to its present length, the reconstructions coincide fully with the spreading and geological fit data. This coincidence does not only apply to the development of the passive-margined oceans, but also to the Pacific, despite all the subduction of crust which has occurred at its active margins. It is not the intention to discuss here the possible causes of global expansion, nor to consider the physical, aqueous and atmospheric problems which arise if one accepts such a concept. The possible answers to some of these problems have been speculated upon elsewhere (Owen 1981). This atlas has been concerned with the testing of the field data and the spherical geometric implications which arise from that data. In the earlier decades of this century, the concept of continental displacement put forward by Alfred Wegener, was ridiculed by most of the geologists and geophysicists of the time. Even when Carey (1958) produced convincing geological evidence for continental displacement which led to the general acceptance of the concept by the mid-1960s, there were no ocean-floor spreading data available to prove it. The ocean-floor spreading data are now to hand, they do not support the concept of a constant modern dimensions Earth nor the concept of plate tectonics as it is generally accepted at the present time. The spreading data do support Earth expansion, yet the concept is not widely accepted!

SECTION 7
World outline maps

MAPS 64–76

Two series of world outline maps are given here for palaeogeographic, palaeobiogeographic and palaeoclimatological studies. These employ Winkel's third ('Tripel') projection (1921) which is, perhaps, the best graticule ever devised to illustrate the physical geography of the Earth, with a minimum of distortion commensurate with a single uninterrupted elliptical map. In order to provide a 'balanced' geographical representation, the 10° east meridian is taken as the prime meridian. Both series of maps have a common integrated enumeration and are spaced at approximately 30 Ma intervals. One series assumes an Earth of constant modern dimensions throughout the Mesozoic and Cenozoic, with the nearest possible fit of the crustal data. The other series assumes an expanding Earth which agrees fully with the oceanic crustal growth illustrated in this atlas. The two series of maps are arranged to run concurrently, with the constant modern dimensions map printed on the left-hand page and its isochronous expanding Earth alternative map printed on the right-hand page. A direct comparison of the two reconstructions, therefore, can easily be made.

The orientation of the Earth's crustal shell to the co-ordinate graticule in each reconstruction backward in time, reflects the relative motion of the continents as the oceanic crust has developed and disrupted the original Pangaea supercontinent. In those maps which assume an Earth of constant modern dimensions, the co-ordinate graticule coincides exactly with the Earth's modern graticule (Map 64). In the case of the expanding Earth maps, only the north co-ordinate pole and the vector of the 0° meridian coincide with their modern counterparts. This method was devised in order to provide common reference graticules by which the relative motions of the continents could be determined accurately. It should not be assumed, therefore, in either series, that the axis of rotation of the Earth at the time of each reconstruction passes through the north and south co-ordinate poles of the map, nor that the geographic equator of the time corresponds to the co-ordinate equator of the map. Such orientations will be determined separately for each period of time on whatever diagnostic data happen to be available.

References

Allen, P. 1976. Wealden of the Weald: a new model. *Proc. Geol. Ass.* London **86**, 389–437, plates 1–3.

Anderson, R. N., Clague, D. A., Klitgord, K. D., Marshall, M. & Nishimori, R. K. 1975. Magnetic and Petrologic Variations along the Galapagos Spreading Center and their relation to the Galapagos melting anomaly. *Bull. Geol. Soc. Amer.* Boulder **86**, 683–94.

Anderson, R. N., Moore, G. F., Schilt, S. S., Cardwell, R. C., Tréhu, A. & Vacquier, V. 1976. Heat flow near a fossil ridge on the north flank of the Galapagos spreading center. *J. Geophys. Res.* Baltimore **81**, 1828–38.

Atwater, T. & Menard, H. W. 1970. Magnetic lineations in the northeast Pacific. *Earth Planet. Sci. Lett.* Amsterdam **7**, 445–50.

Barker, P. F. 1970. Plate tectonics of the Scotia Sea region. *Nature*, London **228**, 1293–6.

Barker, P. F. 1972a. A spreading centre in the east Scotia Sea. *Earth Planet. Sci. Lett.*, Amsterdam **15**, 123–32.

Barker, P. F. 1972b. Magnetic lineations in the Scotia Sea. In Adie, R. J. (Ed.), *Antarctic Geology and Geophysics. Int. Union Geol. Sci.* Oslo **B.1**, 17–26.

Barnett, C. H. 1962. A suggested reconstruction of the land masses of the Earth as a complete crust. *Nature*, London **195**, 447–8.

Barrett, D. L. & Keen, C. E. 1976. Mesozoic magnetic lineations, the magnetic quiet zone, and sea floor spreading in the northwest Atlantic. *J. Geophys. Res.* Baltimore **81**, 4875–84.

Ben-Avraham, Z., Bowin, C. & Segawa, J. 1972. An extinct spreading centre in the Philippine Sea. *Nature*, London **240**, 453–5.

Ben-Avraham, Z. & Uyeda, S. 1973. The evolution of the China Basin and the Mesozoic palaeogeography of Borneo. *Earth Planet. Sci. Lett.* Amsterdam **18**, 365–76.

Bergh, H. W. 1977. Mesozoic sea-floor off Dronning Maud Land, Antarctica. *Nature*, London **269**, 686–7.

Bergh, H. W. & Barrett, D. M. 1980. Agulhas basin magnetic bight. *Nature*, London **287**, 591–5.

Bergh, H. W. & Norton, I. O. 1976. Prince Edward fracture zone and the evolution of the Mozambique Basin. *J. Geophys. Res.* Baltimore **81**, 5221–39.

Bowin, C., Purdy, G. M., Johnston, C., Shor, G., Lawver, L., Hartano, H. M. S. & Jezek, P. 1980. Arc-continent collision in Banda Sea region. *Bull. Am. Ass. Petrol. Geol.* Tulsa **64**, 868–915.

Bracey, D. R. 1975. Reconnaissance geophysical survey of the Caroline Basin. *Bull. Geol. Soc. Amer.* Boulder **86**, 775–84.

Cande, S. C. & Kristoffersen, Y. 1977. Late Cretaceous magnetic anomalies in the North Atlantic. *Earth Planet. Sci. Lett.* Amsterdam **35**, 215–24.

Cande, S. C., Larson, R. L. & La Brecque, J. L. 1978. Magnetic lineations in the Pacific Jurassic quiet zone. *Earth Planet. Sci. Lett.* Amsterdam **41**, 434–40.

Carey, S. Warren 1958. The tectonic approach to continental drift. In Carey S. Warren (Ed.), *Continental Drift – A Symposium*. University of Tasmania, Hobart 177–355, re-printed 1959.

Carey, S. Warren 1970. Australia, New Guinea and Melanesia in the current revolution in concepts of the evolution of the Earth. *Search* Sydney **1**, 178–89.

Carey, S. Warren 1975. The expanding Earth – an essay review. *Earth Sci. Rev.* Amsterdam **11**, 105–43.

Carey, S. Warren 1976. *The Expanding Earth* Developments in Geotectonics **10**, Elsevier, Amsterdam 488pp.

Carey, S. Warren (Ed.) 1983. *The Expanding Earth. A Symposium: Sydney 1981*. University of Tasmania, Hobart i–ix, 1–423.

Christoffel, D. A. & Falconer, R. K. H. 1972. Marine magnetic measurements in the southwest Pacific Ocean and the identification of new tectonic features. *Antarctic Res. Ser. Washington* **19**, 197–209.

Christoffel, D. A. & Ross, D. I. 1970. A fracture zone in the south-west Pacific Basin south of New Zealand and its implications for sea floor spreading. *Earth Planet. Sci. Lett.* Amsterdam **8**, 125–30.

Cocks, L. R. M. (Ed.) 1981. *The Evolving Earth*. British Museum (Natural History) & Cambridge University Press, London & Cambridge i–viii, 1–264.

Cooper, A. K., Scholl, D. W. & Marlow, M. S. 1976. Plate tectonic model for the evolution of the eastern Bering Sea basin. *Bull. Geol. Soc. Amer.* Boulder **87**, 1119–26.

Creer, K. M. 1965. An expanding Earth? *Nature, London* **205**, 539–44.

Dickson, G. O., Pitman, W. C. III & Heirtzler, J. R. 1968. Magnetic anomalies in the South Atlantic and ocean floor spreading. *J. Geophys. Res.* Baltimore **73**, 2087–100.

Egyed, L. 1957. A new dynamic conception of the internal constitution of the Earth. *Geol. Rdsch.* Stuttgart **46**, 101–21.

Elvers, D., Srivastava, S. P., Potter, K., Morley, J. & Sdidel, D. 1973. Asymmetric spreading across the Juan de Fuca and Gorda Rises as obtained from a detailed magnetic survey. *Earth Planet. Sci. Lett.* Amsterdam **20**, 211–19.

Falconer, R. K. H. 1972. The Indian–Antarctic–Pacific triple junction. *Earth Planet. Sci. Lett.* Amsterdam **17**, 151–8.

Firstbrook, P. L., Funnell, B. M., Hurley, A. M. & Smith, A. G. 1980. *Paleoceanic reconstructions 160–0 Ma*. University of California, Scripps Institution of Oceanography 41 pp.

Hallam, A. 1976. How closely did the continents fit together? *Nature, London* **262**, 94–5.

Haller, J. 1969. Tectonics and neotectonics in East Greenland – review bearing on the Drift concept. In North Atlantic – Geology and Continental Drift, a symposium. *Mem. Amer. Ass. Petrol. Geol.* Tulsa **12**, 852–8.

Halm, J. K. E. 1935. An astronomical aspect of the evolution of the Earth. *J. Astr. Soc. S. Afr.* **1**, 1–28.

Handschumacher, D. W. 1976. 16. Post-Eocene plate tectonics of the eastern Pacific. In Sutton, G. H. *et al.* (Eds.), The Geophysics of the Pacific Ocean Basin and its margin. *Geophys. Monogr. Am. Geophys. Union* **19**, 177–202.

Hayes, D. E. 1972. Introduction: marine geophysics of the southeast Indian Ocean. *Antarctic Res. Ser. Washington* **19**, 119–24.

Hayes, D. E. & Pitman, W. C. III 1970. Magnetic lineations in the North Pacific. In Hays, J. D. (Ed.), Geological Investigations of the North Pacific. *Mem. Geol. Soc. Am.* Boulder **126**, 291–314, plates 1–3.

Hayes, D. E. & Rabinowitz, P. D. 1975. Mesozoic magnetic lineations and the magnetic quiet zone off northwest Africa. *Earth Planet. Sci. Lett.* Amsterdam 28, 105–15.

Hayes, D. E. & Ringis, J. 1973. Seafloor spreading in the Tasman Sea. *Nature, London* 243, 454–8.

Hayes, D. E. & Taylor, B. 1978. Tectonics. In Hayes, D. E. *et al.*, *A geophysical Atlas of the East and Southeast Asian Seas.* Geological Society of America.

Heirtzler, J. R., Cameron, P., Cook, P. J., Powell, T., Roeser, H. A., Sukardi, S. & Veevers, J. J. 1978. The Argo abyssal plain. *Earth Planet. Sci. Lett.* Amsterdam 41, 21–31.

Herron, E. M. 1971. Crustal plates and sea-floor spreading in the southeastern Pacific. *Antarctic Res. Ser. Washington* 15, 229–237.

Herron, E. M. 1972. Sea-floor spreading and the Cenozoic history of the East Central Pacific. *Bull. Geol. Soc. Am.* Boulder 83, 1671–91.

Herron, E. M., Dewey, J. F. & Pitman, W. C. III 1974. Plate tectonics model for the evolution of the Arctic. *Geology.* Boulder 2, 377–80.

Herron, E. M. & Tucholke, B. E. 1976. Sea-floor magnetic patterns and basement structure in the south-eastern Pacific. In Hollister, C. D., Craddock, G. *et al.*, *Initial Rep. Deep Sea Drilling Proj.* Washington XXXV, 263–78.

Hey, R. 1977. Tectonic evolution of the Cocos–Nazca spreading center. *Bull. Geol. Soc. Am.* Boulder 88, 1404–20.

Hey, R., Johnson, G. L. & Lowrie, A. 1977. Recent plate motions in the Galapagos area. *Bull. Geol. Soc. Am.* Boulder 88, 1385–403.

Hilde, T. W. C., Isezaki, N. & Wageman, J. M. 1976. 18. Mesozoic sea-floor spreading in the North Pacific. In Sutton, G. H. *et al.* (Eds.), The Geophysics of the Pacific Ocean Basin and its margin. *Geophys. Monogr. Am. Geophys. Union* 19, 205–26.

Hilgenberg, O. 1933. *Vom Wachsenden Erdball.* Berlin 50pp.

Hussong, D. M., Wipperman, L. K. & Kroenke, L. W. 1979. The crustal structure of the Ontong Java and Manihiki oceanic plateaus. *J. Geophys. Res.* Baltimore 84, 6003–10.

Jackson, H. R., Keen, C. E. & Falconer, R. K. H. 1979. New geophysical evidence for sea-floor spreading in central Baffin Bay. *Can. J. Earth Sci.* Ottawa 16, 2122–35.

Jeletzky, J. A. 1980. New or formerly poorly known, biochronologically and palaeobiogeographically important Gastroplitinid and Cleoniceratinid (Ammonitida) taxa from middle Albian rocks of mid-Western and Arctic Canada. *Geol. Surv. Can. Pap.* Ottawa 79–22, i–viii, 1–63, plates 1–10.

Johnson, G. L. & Vogt, P. R. 1973. Marine geology of Atlantic Ocean north of the Arctic circle. *Mem. Am. Ass. Petrol. Geol.* Tulsa 19, 161–70.

Jordan, P. 1971 (1966). *The expanding Earth* (transl. A. Beer from the original German edition 1966). Pergamon Press, Oxford 202pp.

Keen, C. E., Hall, B. R. & Sullivan, K. D. 1977. Mesozoic evolution of the Newfoundland Basin. *Earth Planet. Sci. Lett.* Amsterdam 37, 307–20.

Kobayashi, K. & Isezaki, N. 1976. 20. Magnetic anomalies in the Sea of Japan and the Shikoku Basin: possible tectonic implications. In Sutton, G. H. *et al.* (Eds.), The Geophysics of the Pacific Ocean Basin and its margin. *Geophys. Monogr. Am. Geophys. Union* 19, 235–51.

Kristoffersen, Y. 1978. Sea-floor spreading and the early opening of the North Atlantic. *Earth Planet. Sci. Lett.* Amsterdam 38, 273–90.

Kristoffersen, Y. & Talwani, M. 1977. Extinct triple junction south of Greenland and the Tertiary motion of Greenland relative to North America. *Bull. Geol. Soc. Amer.* Boulder 88, 1037–49.

Kumar, N. & Embley, R. W. 1977. Evolution and origin of the Ceará rise: an aseismic rise in the western equatorial Atlantic. *Bull. Geol. Soc. Amer.* Boulder 88, 683–94.

La Brecque, J. L. & Hayes, D. E. 1979. Seafloor spreading history of the Agulhas basin. *Earth Planet. Sci. Lett.* Amsterdam 45, 411–28.

La Brecque, J. L., Kent, D. V. & Cande, S. C. 1977. Revised magnetic polarity time scale for late Cretaceous and Cenozoic time. *Geology* Boulder 5, 330–5.

Ladd, J. W., Dickson, G. O. & Pitman, W. C. III 1973. The age of the South Atlantic. In Nairn, A. E. M. & Stehli, F. G. (Eds.), *The Ocean Basins and Margins. 1. The South Atlantic.* Plenum Press, New York, London 555–73.

Larson, R. L. 1975. Jurassic sea-floor spreading in the eastern Indian Ocean. *Geology* Boulder 3, 69–71.

Larson, R. L. 1977. Early Cretaceous breakup of Gondwanaland off western Australia. *Geology* Boulder 5, 57–60.

Larson, R. L., Carpenter, G. B. & Diebold, J. B. 1978. A geophysical study of the Wharton Basin near the Investigator fracture zone. *J. Geophys. Res.* Baltimore 83, 773–82.

Larson, R. L. & Chase, C. G. 1972. Late Mesozoic evolution of the western Pacific Ocean. *Bull. Geol. Soc. Am.* Boulder 83, 3627–44.

Larson, R. L. & Hilde, T. W. C. 1975. A revised time scale of magnetic reversals for the early Cretaceous and late Jurassic. *J. Geophys. Res.* Baltimore 80, 2586–94.

Larson, R. L. & Ladd, J. W. 1973. Evidence for the opening of the South Atlantic in the early Cretaceous. *Nature, London* 246, 209–12.

Larson, R. L. & Pitman, W. C. III 1972. World-wide correlation of Mesozoic magnetic anomalies, and its implications. *Bull. Geol. Soc. Amer.* Boulder 83, 3645–62.

Lattimore, R. K., Rona, P. A. & De Wald, O. E. 1974. Magnetic anomaly sequence in the central north Atlantic. *J. Geophys. Res.* Baltimore 79, 1207–9.

Laughton, A. S. 1971. South Labrador Sea and the evolution of the North Atlantic. *Nature, London* 232, 612–17.

Laughton, A. S. 1972. The southern Labrador Sea – a key to the Mesozoic and early Tertiary evolution of the North Atlantic. In Laughton, A. S., Berggren, W. A. *et al.*, *Initial Rep. Deep Sea Drilling Proj.* Washington 12, 1155–79.

Le Pichon, X. & Fox, P. J. 1971. Marginal offsets, fracture zones, and the early opening of the North Atlantic. *J. Geophys. Res.* Baltimore 76, 6294–308.

Le Pichon, X. & Hayes, D. E. 1971. Marginal offsets, fracture zones, and the early opening of the South Atlantic. *J. Geophys. Res.* Baltimore 76, 6283–93.

Lewis, C. & Campbell, J. D. (Eds.) 1951. *The Oxford Atlas.* Oxford University Press pp. 1–8, maps 9–96 & I–XXVI, pp. 1–90.

Lonsdale, P. & Klitgord, K. D. 1978. Structure and tectonic history of the eastern Panama Basin. *Bull. Geol. Soc. Amer.* Boulder 89, 981–99.

Louden, K. E. 1976. 21. Magnetic anomalies in the West Philippine Basin. In Sutton, G. H. *et al.* (Eds.), The Geophysics of the Pacific Ocean Basin and its margin. *Geophys. Monogr. Am. Geophys. Union* 19, 253–67.

Louden, K. E. 1977. Paleomagnetism of DSDP sediments, phase shifting of magnetic anomalies, and rotations of the West Philippine Basin. *J. Geophys. Res.* Baltimore 82, 2989–3002.

Luyendyk, B. P., Bryan, W. B. & Jezek, P. A. 1974. Shallow structure of the New Hebrides Island arc. *Bull. Geol. Soc. Amer.* Boulder 85, 1287–300.

Luyendyk, B. P., Macdonald, K. C. & Bryan, W. B. 1973. Rifting history of the Woodlark Basin in the southwest Pacific. *Bull. Geol. Soc. Amer.* Boulder 84, 1125–33.

McKenzie, D. P. & Sclater, J. G. 1971. The evolution of the Indian Ocean since the late Cretaceous. *Geophys. J. R. Astr. Soc.* Oxford & Edinburgh 25, 437–528.

Malahoff, A. & Handschumacher, D. W. 1971. Magnetic anomalies south of the Murray fracture zone: new evidence for a secondary sea-floor spreading center and strike-slip movement. *J. Geophys. Res.* Baltimore 76, 6265–75.

Mammerickx, J., Anderson, R. N., Menard, H. W. & Smith, S. M. 1975. Morphology and tectonic evolution of the east-central Pacific. *Bull. Geol. Soc. Amer.* Boulder 86, 111–17.

Markl, R. G. 1974. Evidence for the breakup of eastern Gondwanaland by the early Cretaceous. *Nature, London* 251, 196–200.

Markl, R. G. 1978. Basement morphology and refit geometry near the former junction of India, Australia, and Antarctica. *Earth Planet. Sci. Lett.* Amsterdam 39, 211–25.

Mascle, J. & Phillips, J. D. 1972. Magnetic smooth zones in the South Atlantic. *Nature, London* 240, 80–4.

Molnar, P., Atwater, T., Mammerickx, J. & Smith, S. M. 1975. Magnetic anomalies, bathymetry and the tectonic evolution of the south Pacific since the late Cretaceous. *Geophys. J. R. Astr. Soc.* Oxford & Edinburgh 40, 383–420.

Monmonier, M. S. 1982. *Computer-assisted cartography: principles and prospects.* Prentice-Hall, Englewood Cliffs, N.J. 214pp.

Murakami, F., Tamaki, K. & Nishimura, K. 1977. V. Geomagnetic survey. In Honza, E. (Ed.), Geological investigation of Japan and southern Kurile Trench and slope areas GH 76-2 Cruise April–June 1976. *Cruise Rep. Geol. Surv. Japan* 7, 43–9.

Nairn, A. E. M., Stehli, F. G. et al. (Eds.), 1973–81. *The Ocean Basins and Margins*. Plenum Press, New York & London 7 vols (5 published).

Norton, I. O. & Sclater, J. G. 1979. A model for the evolution of the Indian Ocean and the breakup of Gondwanaland. *J. Geophys. Res.* Baltimore 84, 6803–30.

Olivet, J-L., Le Pichon, X., Monti, S. & Sichler, B. 1974. Charlie–Gibbs fracture zone. *J. Geophys. Res.* Baltimore 79, 2059–72.

Owen, H. G. 1971. The stratigraphy of the Gault in the Thames Estuary and its bearing on the Mesozoic tectonic history of the area. *Proc. Geol. Ass.* London 82, 187–207.

Owen, H. G. 1973. Ammonite faunal provinces in the middle and upper Albian and their palaeogeographical significance. In Casey, R. & Rawson, P. F. (Eds.), The Boreal Lower Cretaceous. *Geol. Jl.* Spec. Issue 5, 145–54.

Owen, H. G. 1976. Continental displacement and expansion of the Earth during the Mesozoic and Cenozoic. *Phil. Trans. R. Soc. Lond.* London A281, 223–91.

Owen, H. G. 1981. Constant dimensions or an expanding Earth. In Cocks, L. R. M. (Ed.), *The Evolving Earth*. British Museum (Natural History) and Cambridge University Press, London & Cambridge 179–92.

Owen, H. G. 1983. Ocean-floor spreading evidence of global expansion. In Carey, S. Warren (Ed.), *The Expanding Earth. A Symposium: Sydney 1981*. University of Tasmania, Hobart 31–58.

Peter, G., Lattimore, R. K., De Wald, O. E. & Merrill, G. 1973. Development of the mid-Atlantic ridge east of the Lesser Antilles island arc. *Nature, London (Phys. Sci.)* 245, 129–31.

Phillips, J. D., Fleming, H. S., Feden, R. H., King, W. E. & Perry, R. K. 1975. Aeromagnetic study of the mid-Atlantic ridge near the Oceanographer fracture zone. *Bull. Geol. Soc. Amer.* Boulder 86, 1348–57.

Pitman, W. C. III, Larson, R. L. & Herron, E. M. 1974. *The age of the ocean basins*. Geological Society of America, Boulder.

Pitman, W. C. III, Talwani, M. & Heirtzler, J. R. 1971. Age of the North Atlantic Ocean from magnetic anomalies. *Earth Planet. Sci. Lett.* Amsterdam 11, 195–200.

Purdy, G. M. & Rohr, K. 1970. A geophysical survey within the Mesozoic magnetic anomaly sequence south of Bermuda. *J. Geophys. Res.* Baltimore 84, 5487–96.

Rabinowitz, P. D., Cande, S. C. & Hayes, D. E. 1979. 45. The J-anomaly in the central North Atlantic Ocean. In Tucholke, B. E., Vogt, P. R. et al., *Initial Rep. Deep Sea Drilling Proj.* Washington 43, 879–85.

Rabinowitz, P. D. & La Brecque, J. 1979. The Mesozoic South Atlantic ocean and evolution of its continental margins. *J. Geophys. Res.* Baltimore 84, 5973–6002.

Rabinowitz, P. D. & Purdy, G. M. 1976. The Kane fracture zone in the western central Atlantic Ocean. *Earth Planet. Sci. Lett.* Amsterdam 33, 21–6.

Ramberg, I. B., Gray, D. F. & Raynolds, R. G. H. 1977. Tectonic evolution of the FAMOUS area of the Mid-Atlantic Ridge, lat. 35° 50' to 37° 20' N. *Bull. Geol. Soc. Amer.* Boulder 88, 609–20.

Schlich, R. 1974. 30. Sea floor spreading history and deep-sea drilling results in the Madagascar and Mascarene basins, western Indian Ocean. In Simpson, E. S. W., Schlich, R. et al., *Initial Rep. Deep Sea Drilling Proj.* Washington 25, 663–78.

Sclater, J. G., Bowin, C., Hey, R., Hoskins, H., Peirce, J., Phillips, J. & Tapscott, C. 1976. The Bouvet triple junction. *J. Geophys. Res.* Baltimore 81, 1857–69.

Sclater, J. G. & Fisher, R. L. 1974. Evolution of the east central Indian Ocean, with emphasis on the tectonic setting of the Ninetyeast ridge. *Bull. Geol. Soc. Amer.* Boulder 85, 683–702.

Sclater, J. G. & Klitgord, K. D. 1973. A detailed heat flow, topographic, and magnetic survey across the Galapagos spreading center at 86° W. *J. Geophys. Res.* Baltimore 78, 6951–75.

Sclater, J. G., Luyendyk, B. P. & Meinke, L. 1976. Magnetic lineations in the southern part of the central Indian basin. *Bull. Geol. Soc. Amer.* Boulder 87, 371–8.

Ségoufin, J. 1978. Anomalies magnétiques mésozoïques dans le bassin de Mozambique. *C. R. Hebd. Séanc. Acad. Sci. Paris* 287D, 109–12.

Smith, A. G., Hurley, A. M. & Briden, J. C. 1981. *Phanerozoic Paleocontinental World Maps*. Cambridge University Press 102pp.

Spencer, A. M. (Ed.) 1974. *Mesozoic–Cenozoic orogenic belts. Data for orogenic studies*. The Geological Society Spec. Pub. 4. Scottish Academic Press & The Geological Society, Edinburgh & London i–xvi, 1–809.

Steers, J. A. 1962. *An Introduction to the Study of Map Projections*. 13th Edition. University of London Press 258pp.

Steiner, J. 1977. An expanding Earth on the basis of sea-floor spreading and subduction rates. *Geology*. Boulder 5, 313–18.

Steiner, M. B. 1977. Magnetization of Jurassic red deep-sea sediments in the Atlantic (DSDP Site 105). *Earth Planet. Sci. Lett.* Amsterdam 35, 205–14.

Stöklin, J. 1977. Structural correlation of the Alpine ranges between Iran and Central Asia. *Mém. H. Sér. Soc. Géol. Fr.* Paris 8, 333–53.

Stöklin, J. 1980. Geology of Nepal and its regional frame. *J. Geol. Soc. Lond.* 137, 1–34.

Storetvedt, K. M. 1972. Crustal evolution in the Bay of Biscay. *Earth Planet. Sci. Lett.* Amsterdam 17, 135–41.

Surlyk, F. 1978a. Submarine fan sedimentation along fault scarps on tilted fault blocks (Jurassic–Cretaceous boundary, East Greenland). *Bull. Grøn. Geol. Unders.* Copenhagen 128, 1–108, Appendix, plates 1–9.

Surlyk, F. 1978b. Mesozoic geology and palaeogeography of Hochstetter Forland, East Greenland. *Bull. Geol. Soc. Denmark* Copenhagen 27, 73–87.

Talwani, M. & Eldholm, O. 1977. Evolution of the Norwegian–Greenland Sea. *Bull. Geol. Soc. Amer.* Boulder 88, 969–99.

Tamaki, K., Joshima, M. & Larson, R. L. 1979. Remanent early Cretaceous spreading center in the Central Pacific Basin. *J. Geophys. Res.* Baltimore 84, 4501–10.

Taylor, P. T. & Greenwalt, D. 1976. Geophysical transitions across the northwest Atlantic magnetic quiet-zone border. *Earth Planet. Sci. Lett.* Amsterdam 29, 435–46.

Tobler, W. R. 1973. The hyperelliptical and other new pseudo-cylindrical equal area map projections. *J. Geophys. Res.* Baltimore 78, 1753–9.

Truchan, M. & Larson, R. L. 1973. Tectonic lineaments on the Cocos Plate. *Earth Planet. Sci. Lett.* Amsterdam 17, 426–32.

Van Andel, T. H., Rea, D. K., Von Herzen, R. P. & Hoskins, H. 1973. Ascension fracture zone, Ascension Island, and mid-Atlantic ridge. *Bull. Geol. Soc. Amer.* Boulder 84, 1527–46.

Vink, G. E. 1982. Continental rifting and the implications for plate tectonic reconstructions. *J. Geophys. Res.* Baltimore 87, 10677–88.

Vogt, P. R., Anderson, C. N. & Bracey, D. R. 1971. Mesozoic magnetic anomalies, sea-floor spreading, and geomagnetic reversals in the southwestern North Atlantic. *J. Geophys. Res.* Baltimore 76, 4796–823.

Vogt, P. R. & Avery, O. E. 1974. Detailed magnetic surveys in the northeast Atlantic and Labrador Sea. *J. Geophys. Res.* Baltimore 79, 363–89.

Vogt, P. R. & Byerly, G. R. 1976. Magnetic anomalies and basalt composition in the Juan de Fuca–Gorda Ridge area. *Earth Planet. Sci. Lett.* Amsterdam 33, 185–207.

Vogt, P. R. & De Boer, J. 1976. Morphology, magnetic anomalies and basalt magnetization at the ends of the Galapagos high-amplitude zone. *Earth Planet. Sci. Lett.* Amsterdam 33, 145–63.

Vogt, P. R. & Einwich, A. M. 1979. 43. Magnetic anomalies and sea-floor spreading in the western North Atlantic, and a revised calibration of the Keathley (M) geomagnetic reversal chronology. In Tucholke, B. E., Vogt, P. R. et al., *Initial Rep. Deep Sea Drilling Proj.* Washington 43, 857–76.

Vogt, P. R. & Johnson, G. L. 1971. Cretaceous sea floor spreading in the western North Atlantic. *Nature, London* 234, 22–5.

Vogt, P. R. & Ostenso, N. A. 1970. Magnetic and gravity profiles across the Alpha Cordillera and their relation to Arctic sea-floor spreading. *J. Geophys. Res.* Baltimore 75, 4925–37.

Vogt, P. R., Taylor, P. T., Kovacs, L. C. & Johnson, G. L. 1979. Detailed aeromagnetic investigation of the Arctic Basin. *J. Geophys. Res.* Baltimore 84, 1071–89.

Watts, A. B. & Weissel, J. K. 1975. Tectonic history of the Shikoku
   marginal basin. *Earth Planet. Sci. Lett.* Amsterdam **25**, 239–50.
Weissel, J. K. & Hayes, D. E. 1972. Magnetic anomalies in the
   southeast Indian Ocean. *Antarctic Res. Ser. Washington* **19**,
   165–96.
Weissel, J. K. & Hayes, D. E. 1977. Evolution of the Tasman Sea
   reappraised. *Earth Planet. Sci. Lett.* Amsterdam **36**, 77–84.
Weissel, J. K. & Watts, A. B. 1975. Tectonic complexities in the South
   Fiji marginal basin. *Earth Planet. Sci. Lett.* Amsterdam **28**, 121–6.
Weissel, J. K. & Watts, A. B. 1979. Tectonic evolution of the Coral
   Sea Basin. *J. Geophys. Res.* Baltimore **84**, 4572–82.
Williams, C. A. & McKenzie, D. 1971. The evolution of the north-east
   Atlantic. *Nature, London* **232**, 168–73.
Winkel, O. 1921. Neue Gradnetzkombinationen. *Petermanns Mitt.*
   Gotha **67**, 248–52, plate 24.

Appendix

of some recent papers

The following important papers have been published since the reconstructions mapped in this atlas were prepared. None of them require revisions to be made to the argument for global expansion given here. It should be noted that no additional ocean-floor spreading data were sought after July 1980, until the completion of the maps. Any coincidence of more recent data with the reconstructions here is a function, therefore, of the constraints upon the development of a given area provided by the previously mapped areas surrounding it.

1. Introduction: Briden, Hurley & Smith (1981) have reiterated the view held widely by palaeomagneticists, that the Earth's magnetic field was of the same average dimensions and order of intensity in the past as it is today. This is an assumed working hypothesis that is not a proven fact and, indeed, much of the palaeomagnetic data coincides with the arrangement of the continents during the Mesozoic and Cenozoic in the expanding Earth/continental displacement model described in this atlas. Their maps, while possessing cartographic integrity, do not indicate the ocean-floor spreading data upon which their reconstructions are partly based.

Briden, J. C., Hurley, A. M. & Smith, A. G. 1981. Paleomagnetism and Mesozoic–Cenozoic paleocontinental maps. *J. Geophys. Res.* Washington **86**, 11631–56.

2. Section 1, Maps 4–13: Taylor, Kovacs, Vogt & Johnson (1981) have discussed the history of formation of the Eurasia Basin and of the Canada Basin. If their identifications of the age of the magnetic anomalies in the Canada Basin are correct, the rotation of Alaska away from Arctic Canada to form this basin commenced in the early upper Jurassic (a little before Anomaly M25) and finished at Anomaly M12 (Valanginian). However, this does not agree with the continental palaeomagnetic rotational data from Alaska which indicate a lower Cretaceous age for the opening and formation of this basin, a discrepancy which they acknowledge.

Taylor, P. T., Kovacs, L. C., Vogt, P. R. & Johnson, G. L. 1981. Detailed aeromagnetic investigation of the Arctic Basin, 2. *J. Geophys. Res.* Washington **86**, 6323–33.

3. Section 2, Maps 14, 16–20: Roberts, Masson & Miles (1981) have concluded from the geophysical data that the separation of the Rockall Plateau from the north-west European margin, by ocean-floor spreading, took place in the interval between the Albian and the Maastrichtian (upper Cretaceous).

Roberts, D. G., Masson, D. G. & Miles, P. R. 1981. Age and structure of the southern Rockall Trough: new evidence. *Earth Planet. Sci. Lett.* Amsterdam **52**, 115–28.

4. Section 3: Kellogg (1980) has discussed the evidence of oroclinal bending in the southern region of the West Antarctic Peninsula.

Kellogg, K. S. 1980. Paleomagnetic evidence for oroclinal bending of the southern Antarctic Peninsula. *Bull. Geol. Soc. Amer.* Boulder **91**, 414–20.

5. Section 4, Maps 2, 38–43: Johnson, Powell & Veevers (1980) and Veevers, Powell & Johnson (1980) have discussed the history of the early opening of the Indian Ocean assuming a constant modern dimensions Earth. Their reconstructions of the development of the whole area of the Indian Ocean are cartoons and do not represent correctly projected maps. Their maps of the distribution of magnetic anomalies seen today are of their usual high standard.

Johnson, B. D., Powell, C. McA. & Veevers, J. J. 1980. Early spreading history of the Indian Ocean between India and Australia. *Earth Planet. Sci. Lett.* Amsterdam **47**, 131–43.
Veevers, J. J., Powell, C. McA. & Johnson, B. D. 1980. Seafloor constraints on the reconstruction of Gondwanaland. *Earth Planet. Sci. Lett.* Amsterdam **51**, 435–44.

6. Section 5, Maps 3, 44–53: Farrar & Dixon (1981) have also discussed the role of the Emperor chain, their Emperor fracture zone, in the differential movement of oceanic crust during the later development of the northern Pacific. They arrived at the same conclusion as the present author concerning the relationship of the Great Magnetic Bight to the magnetic bight of the Japanese and Hawaiian plates.

Farrar, E. & Dixon, J. M. 1981. Early Tertiary rupture of the Pacific plate: 1700 km of dextral offset along the Emperor Trough – Line Islands lineament. *Earth Planet. Sci. Lett.* Amsterdam **53**, 307–22.

7. Section 5, Maps 3, 54–7: Cande, Herron & Hall (1982) have provided additional information on the Cenozoic spreading patterns and history of the south-eastern region of the Pacific to the west of the Scotia Sea.

Cande, S. C., Herron, E. M. & Hall, B. R. 1982. The early Cenozoic tectonic history of the southeast Pacific. *Earth Planet. Sci. Lett.* Amsterdam **57**, 63–74.

8. Section 5, Maps 3, 54–7: Malahoff, Feden & Fleming (1982) have described the Oligocene–Recent spreading patterns of the North and South Fiji Basins.

Malahoff, A., Feden, R. H. & Fleming, H. S. 1982. Magnetic anomalies and tectonic fabric of marginal basins north of New Zealand. *J. Geophys. Res.* Washington **87**, 4109–25.

9. Maps 3, 54–7: Mammerickx, Herron & Dorman (1980) have rotated the Marquesas, Austral and Agassiz fracture zones to join up with the Mendāna, Easter Island and Challenger fracture

zones, much as in the reconstructions given in this atlas. Pilger and Handschumacher (1981) have used the evidence of a common hotspot to relocate the Tuamotu and Nazca ridges together, and their subsequent Miocene to Recent displacement is marked by the development of the Easter Island/Sala y Gomez ridge system.

Mammerickx, J., Herron, E. & Dorman, L. 1980. Evidence for two fossil spreading ridges in the southeast Pacific. *Bull. Geol. Soc. Amer.* Boulder **91**, 263–71.

Pilger, R. H. & Handschumacher, D. W. 1981. The fixed-hotspot hypothesis and the origin of the Easter–Sala-y-Gomez–Nazca trace. *Bull. Geol. Soc. Amer.* Boulder **92**, 437–46.

10. Section 5, Maps 3, 44–7: Klitgord & Mammerickx (1982) and Mammerickx & Klitgord (1982) have described in detail the magnetic anomaly patterns in the eastern Pacific from 7° north to 30° north latitudes and east of 127° west longitude. They give a fine picture of the Miocene to Recent spreading patterns which have broken through the older Cenozoic pattern north of the Galapagos spreading zone.

Klitgord, K. D. & Mammerickx, J. 1982. Northern East Pacific Rise: magnetic anomaly and bathymetric framework. *J. Geophys. Res.* Washington **87**, 6725–50.

Mammerickx, J. & Klitgord, K. D. 1982. Northern East Pacific Rise: evolution from 25 m.y. B.P. to the Present. *J. Geophys. Res.* Washington **87**, 6751–9.

Part 2
THE MAPS

Modern Earth

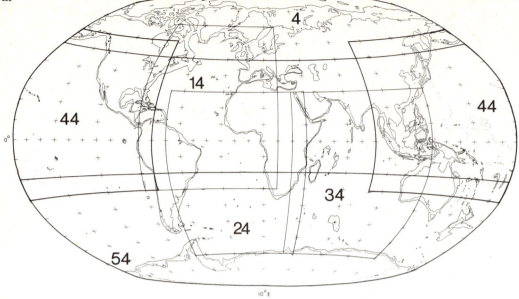

Oligocene (Anomaly 9)
97% of modern diameter

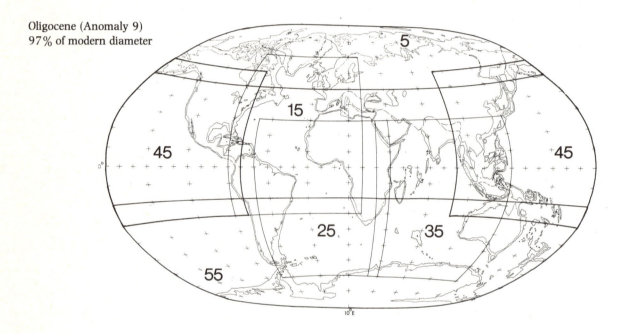

Palaeocene (Anomaly 24)
modern diameter

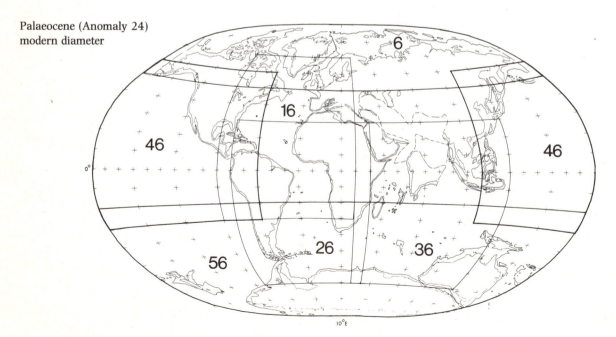

Palaeocene (Anomaly 24)
94% of modern diameter

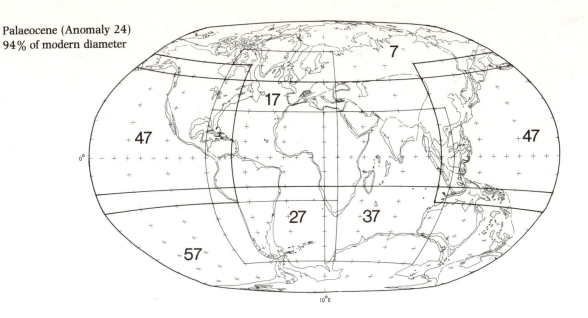

Turonian (90 Ma)
90% of modern diameter

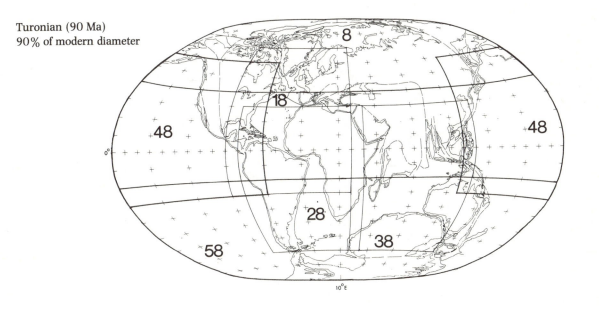

Hauterivian (Anomaly M7)
modern diameter

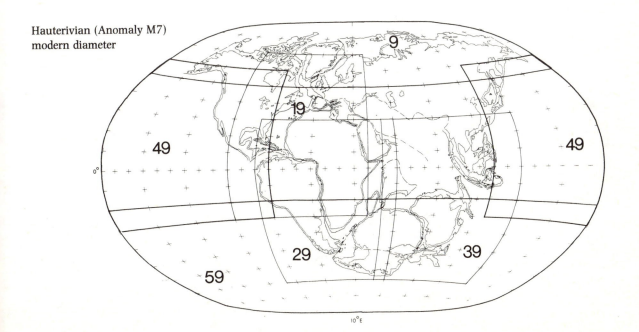

Hauterivian (Anomaly M7)
87% of modern diameter

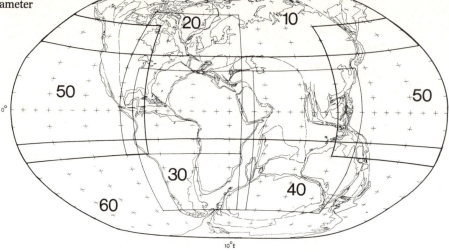

Oxfordian (Anomaly M23)
84% of modern diameter

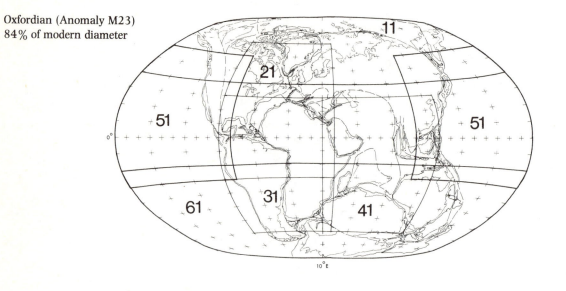

Pangaea (180–200 Ma)
modern diameter

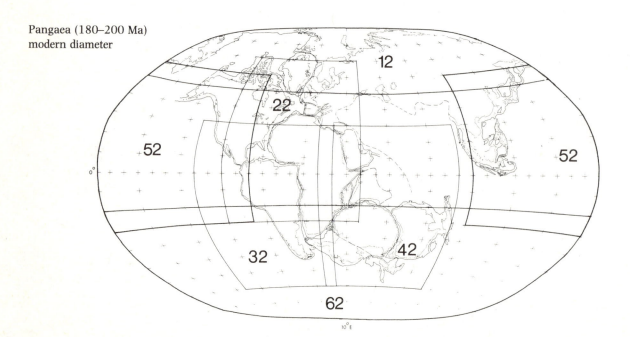

Pangaea (180–200 Ma)
80% of modern diameter

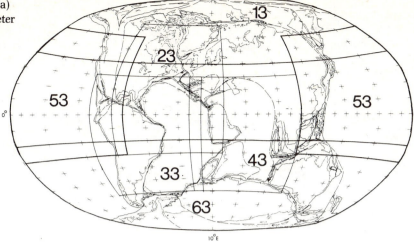

13

23

53   53

33   43

63

LEGEND

  1000m isobath and dotted modern coastline

  major continental wrench fault zones with sense of differential movement

  Tethyan Ocean boundary (constant modern dimensions Earth)

  oceanic fracture zone

  North magnetic pole (constant modern dimensions Earth for comparison with Smith et al.)

  dotted active spreading axis and pecked magnetic normal anomalies

  extinct spreading axis

  Mesozoic magnetic reversed anomalies

magnetic 'quiet zone' boundary

  oceanic deep trench

areas of crustal overlap required in constant modern dimensions Earth reconstructions

  Pacific Ocean modern dimensions Earth reconstructions – oceanic crustal area greater than that required to be subducted in the expanding Earth reconstruction.

Excess oceanic crust (gores etc) required to be eliminated in passive-margin oceans and marginal basins – constant modern dimensions Earth reconstructions.

oceanic crustal area of which, although now subducted, there is firm evidence of its former existence

  Deep Sea Drilling Project borehole site and number

  map projection pole

Map 1

Showing ocean-floor spreading data and Deep Sea
Drilling Project (DSDP) borehole sites (for sources see
text)

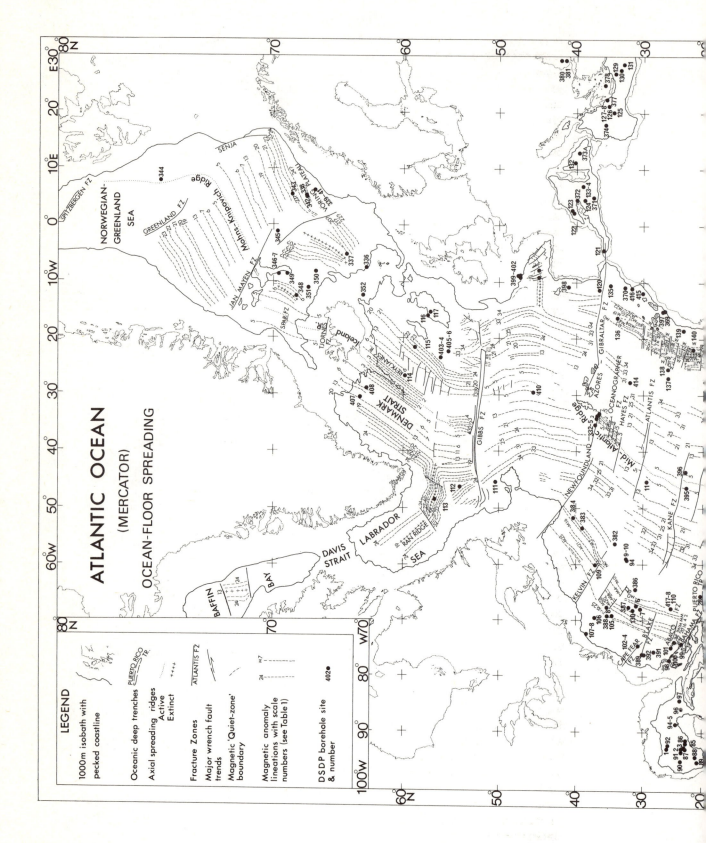

Map 2

Showing ocean-floor
spreading data and Deep Sea
Drilling Project (DSDP)
borehole sites (for sources see
text)

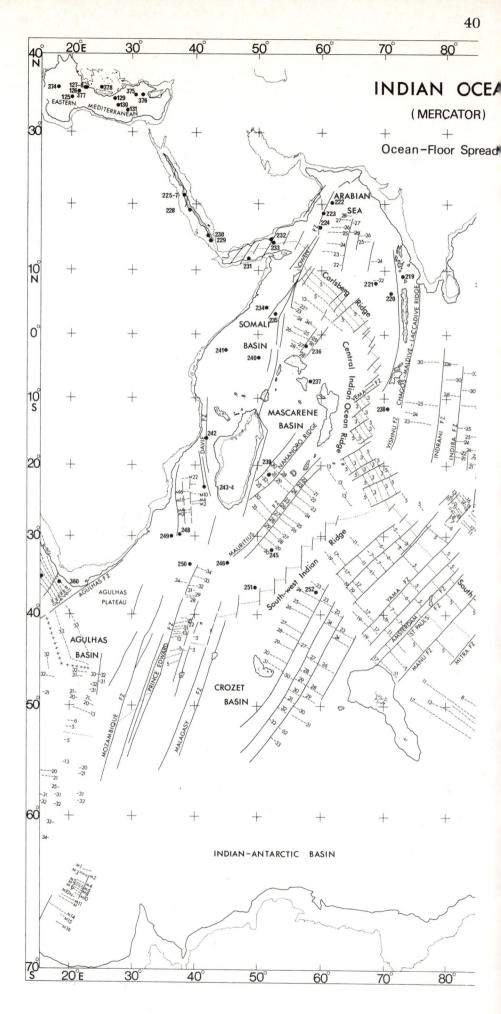

INDIAN OCEAN

(MERCATOR)

Ocean-Floor Spreading

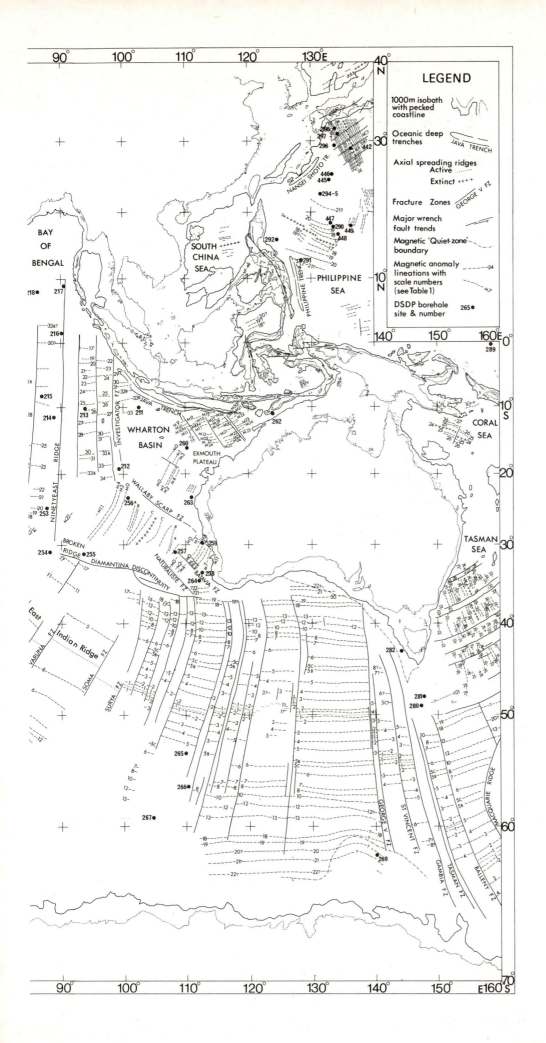

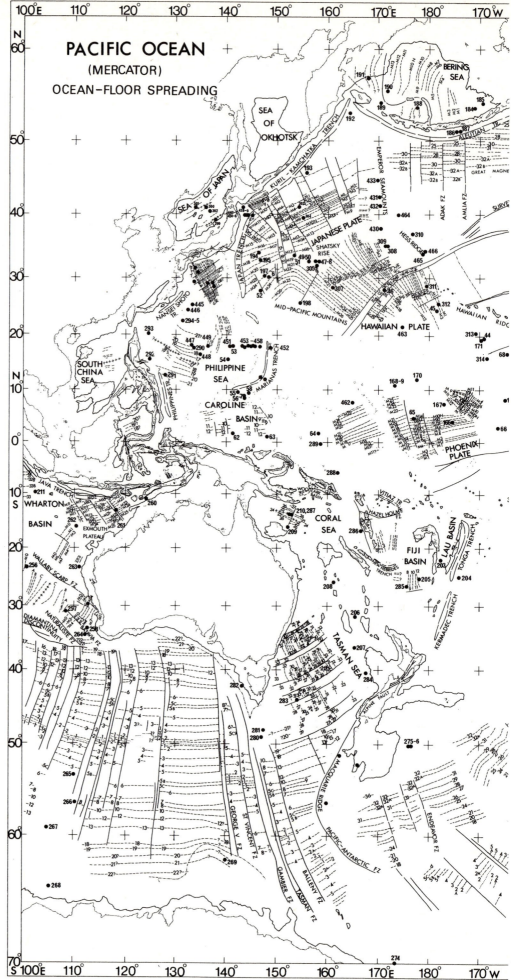

Map 3

Showing ocean-floor
spreading data and Deep
Sea Drilling Project
(DSDP) borehole sites
(for sources see text)

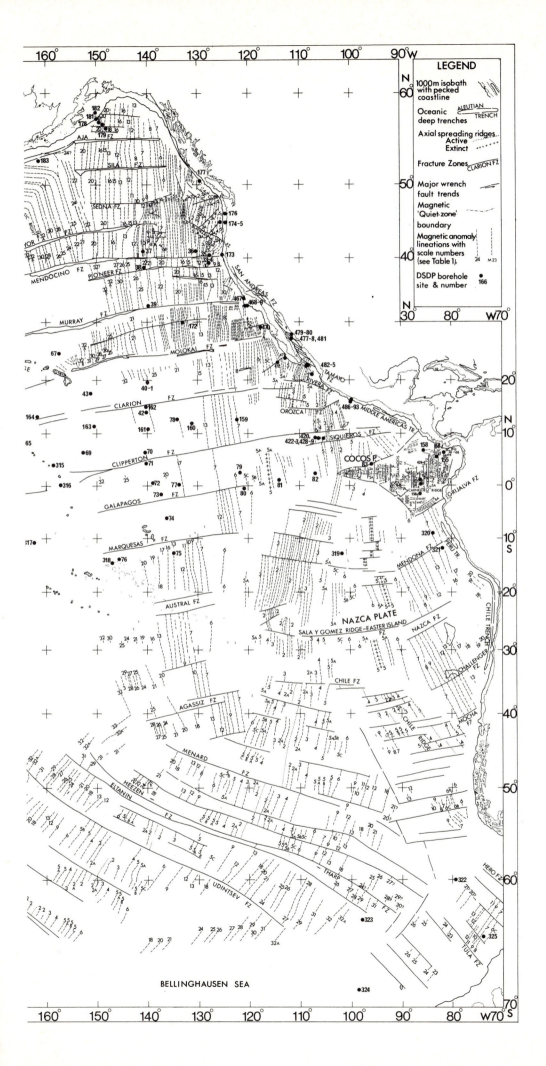

SECTION 1

Boreal region

MAPS 4–13

(azimuthal equidistant projection; polar case)

Map 4

Modern. For sources see text

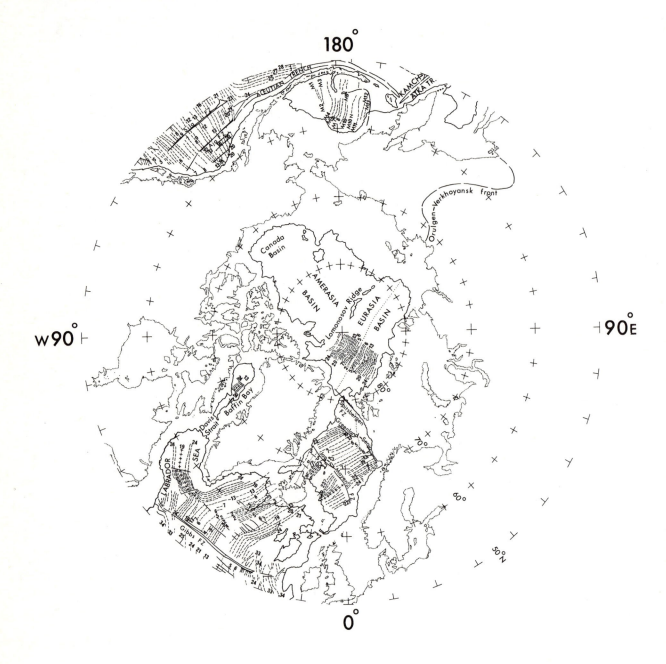

Map 5

Anomaly 9 (29 Ma) Oligocene
ca 97% of modern diameter

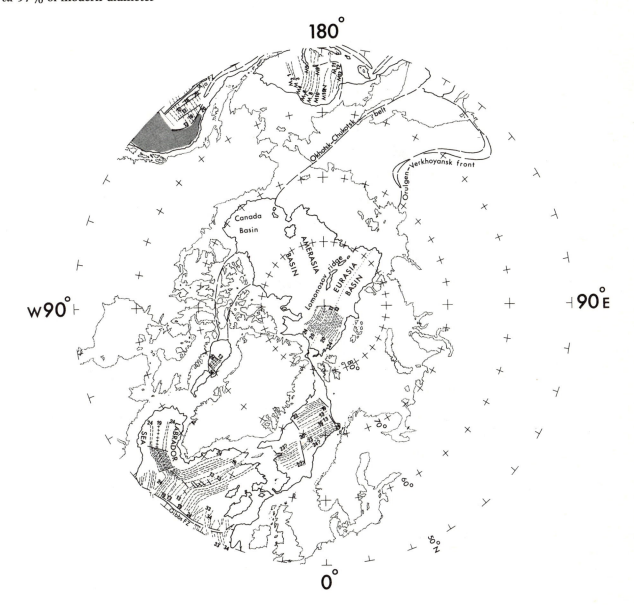

Map 6

Anomaly 24 (56 Ma) Palaeocene
modern dimensions Earth

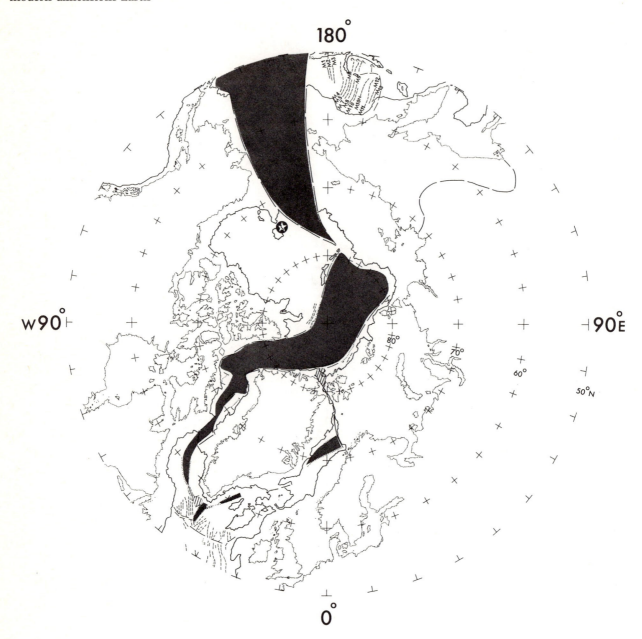

Map 7

Anomaly 24 (56 Ma) Palaeocene
ca 94% of modern diameter

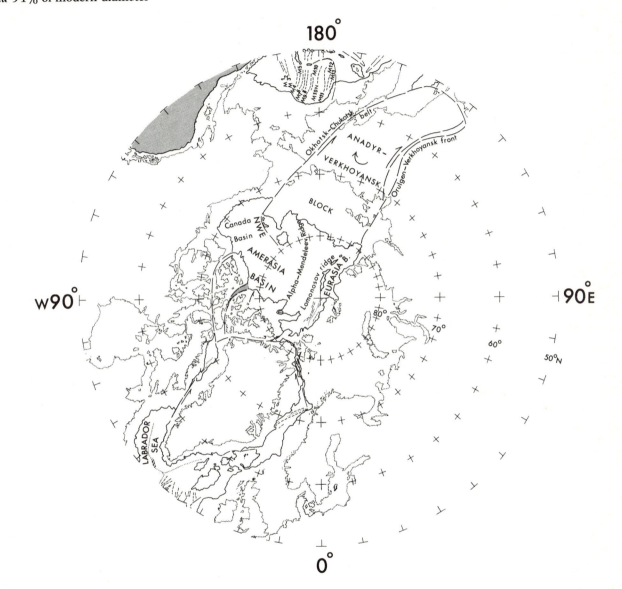

Map 8

Turonian (90 Ma) upper Cretaceous
90% of modern diameter

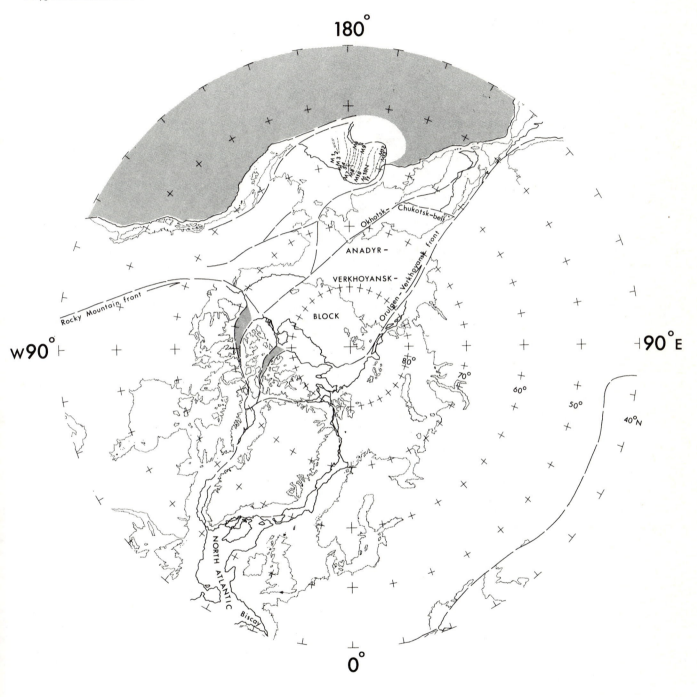

Map 9

Anomaly M7 (120 Ma) Hauterivian
modern dimensions Earth

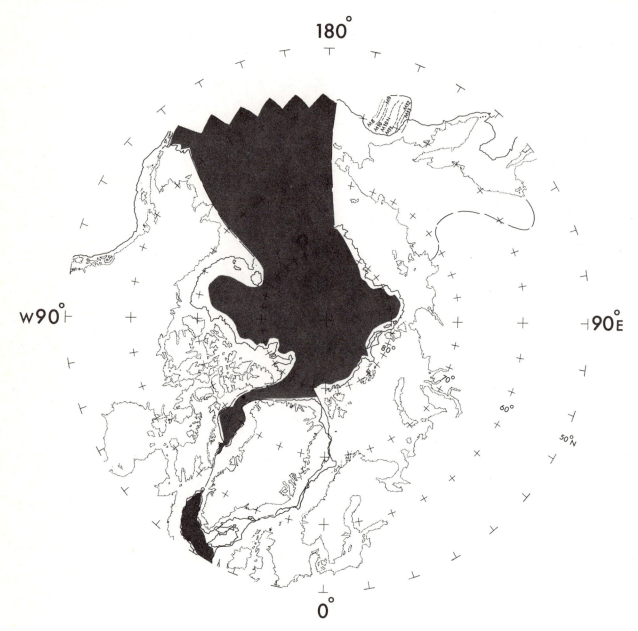

Map 10

Anomaly M7 (120 Ma) Hauterivian
ca 87% of modern diameter

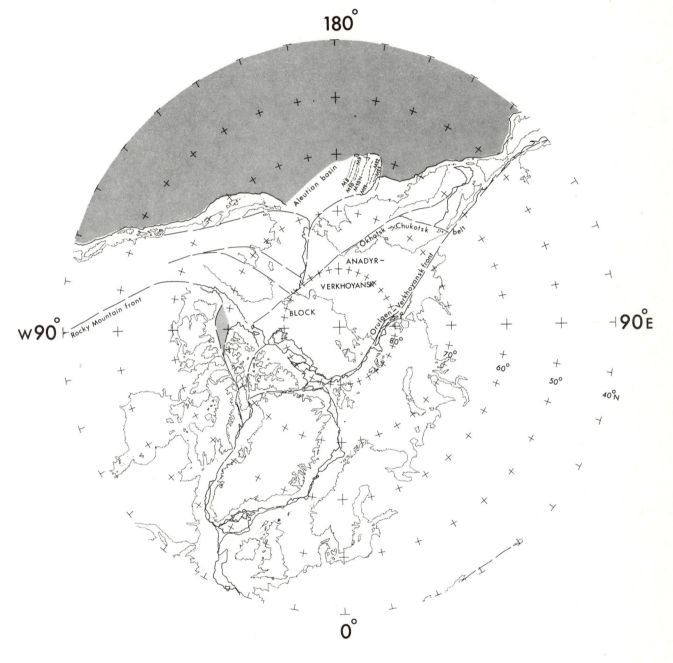

Map 11

Anomaly M23 (146 Ma) Oxfordian
ca 84% of modern diameter

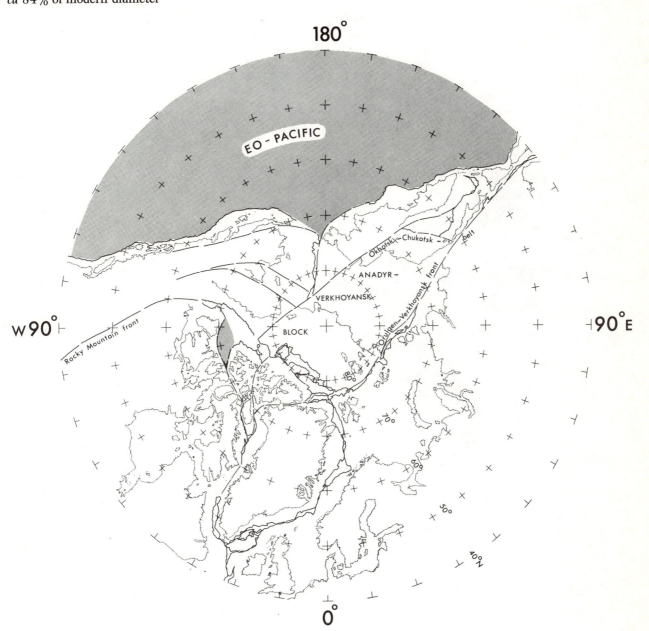

Map 12

Pangaea (180–200 Ma) late Triassic to lower Jurassic
modern dimensions Earth

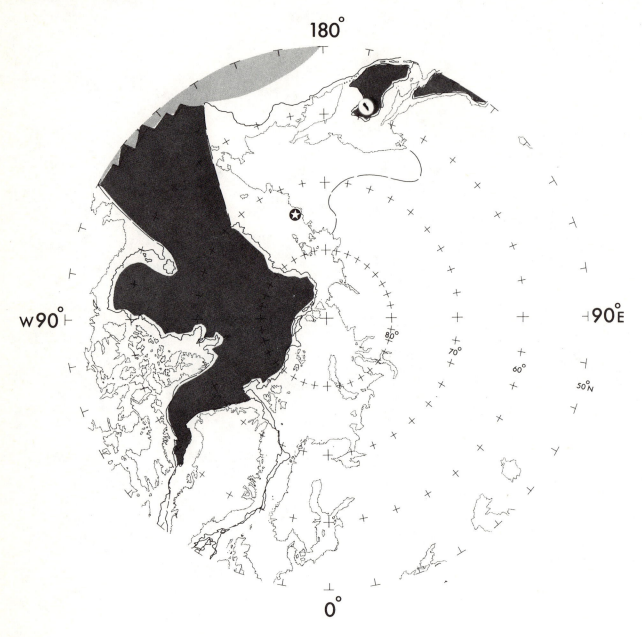

Map 13

Pangaea (180–200 Ma) late Triassic to lower Jurassic
80% of modern diameter

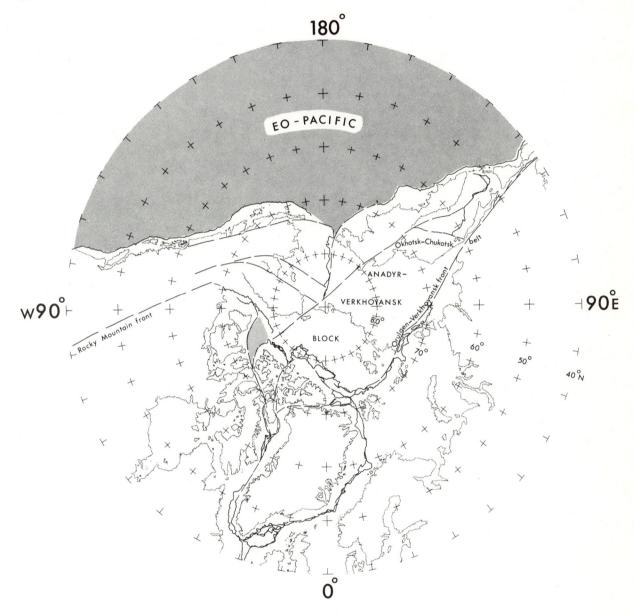

SECTION 2
North Atlantic
MAPS 14–23

(azimuthal equidistant projection; oblique case)

Map 14

Modern
For sources see text
Projection pole 22°N, 30°W

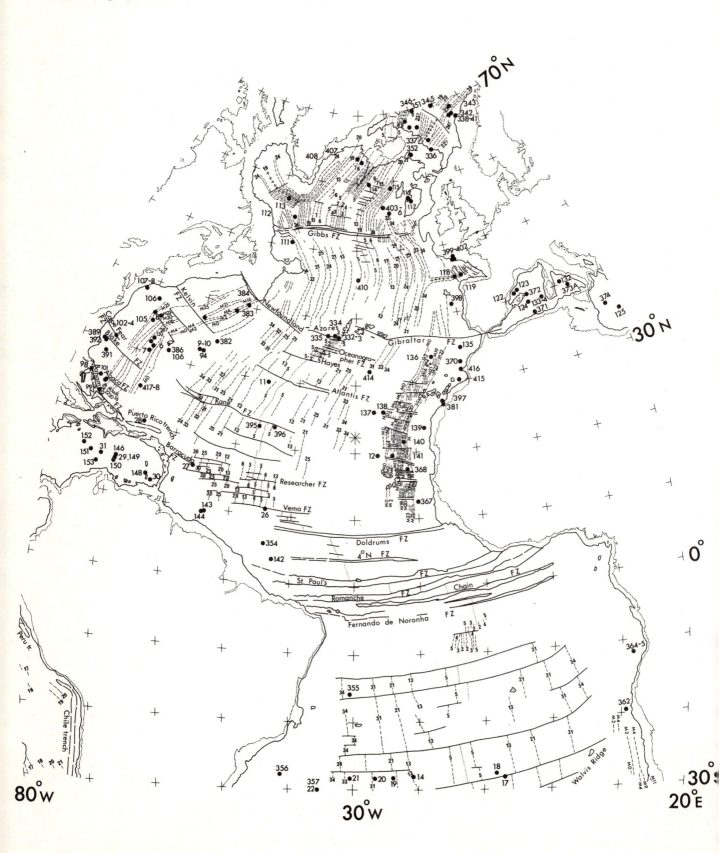

Map 15

Anomaly 9 (29 Ma) Oligocene
ca 97% of modern diameter
Projection pole 22°N, 30°W

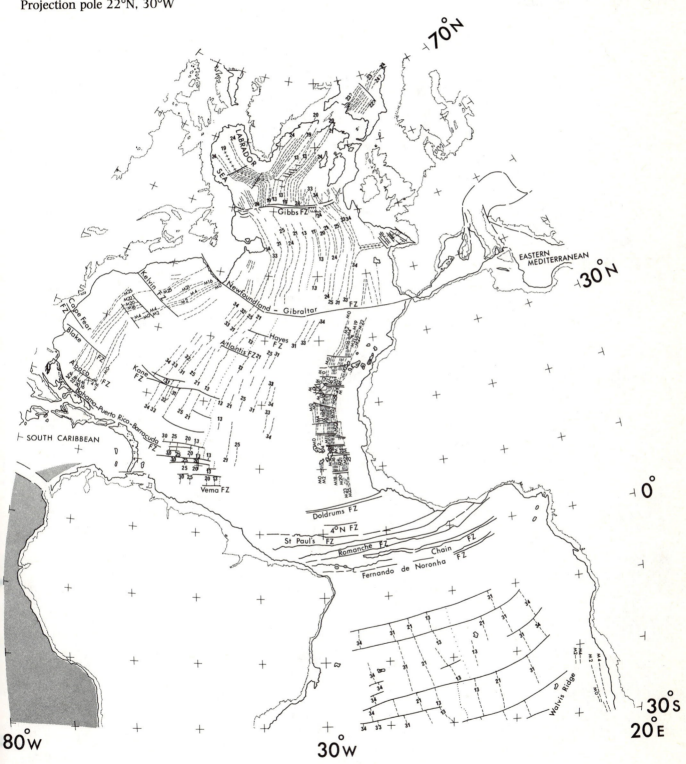

Map 16

Anomaly 24 (56 Ma) Palaeocene
modern dimensions Earth
Projection pole 22°N, 30°W

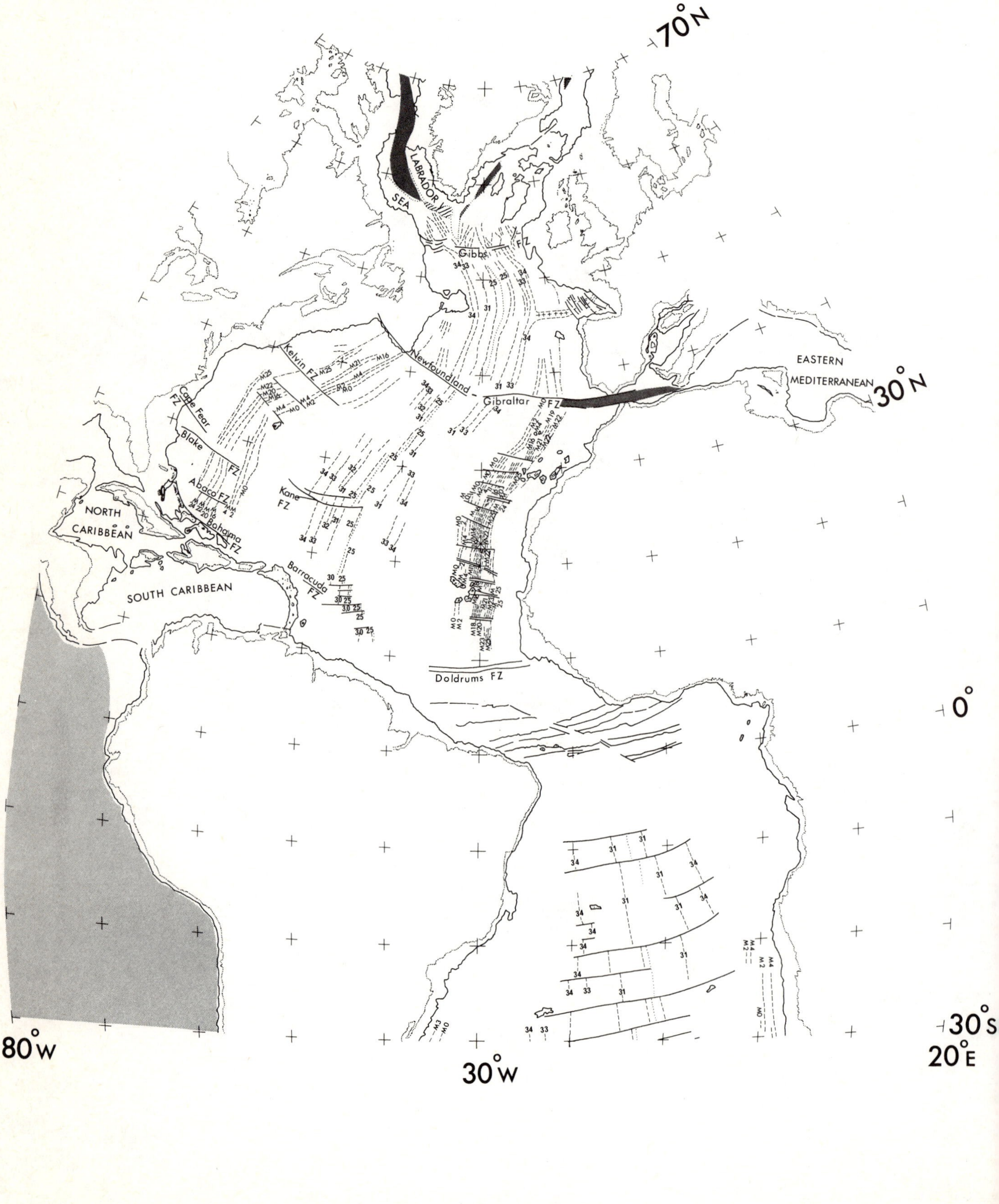

70°N

LABRADOR SEA

Gibbs FZ

25
33
31
34

Kelvin FZ
M25
M21
M16
Newfoundland

M22
M20
M4
M0
M4
M2

Cape Fear FZ

EASTERN MEDITERRANEAN

30°N

Blake FZ

Gibraltar FZ

Abaco FZ

Kane FZ

NORTH CARIBBEAN

Bahama FZ

34 33

33 34

25

34

SOUTH CARIBBEAN

Barracuda FZ

30 25
30 25
25
30 25

M0
M2
M18
M23
M25

Doldrums FZ

0°

31
34
34
31
34
34
34
31
34 33 31

M4
M2
M0

34 33

30°S

80°W

30°W

20°E

Map 17

Anomaly 24 (56 Ma) Palaeocene
ca 94% of modern diameter
Projection pole 22°N, 30°W

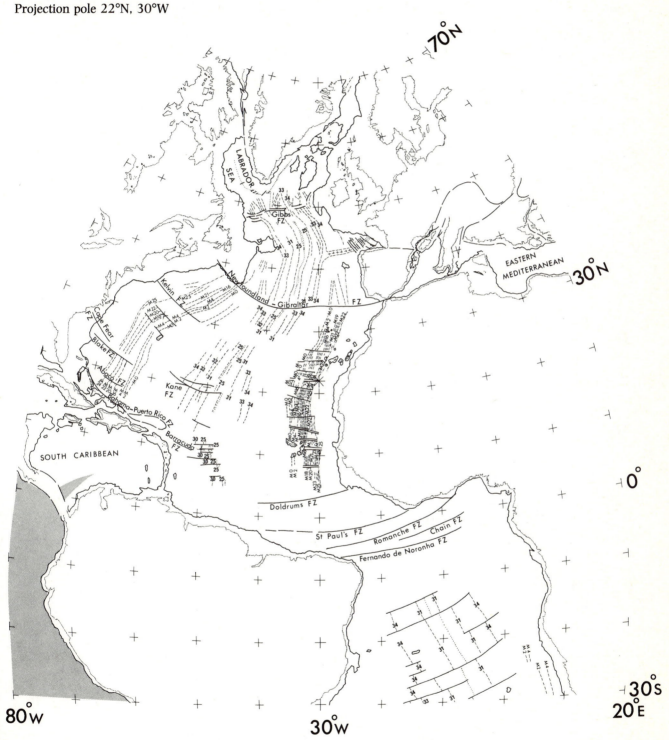

Map 18

Turonian (90 Ma) upper Cretaceous
90% of modern diameter
Projection pole 22°N, 30°W

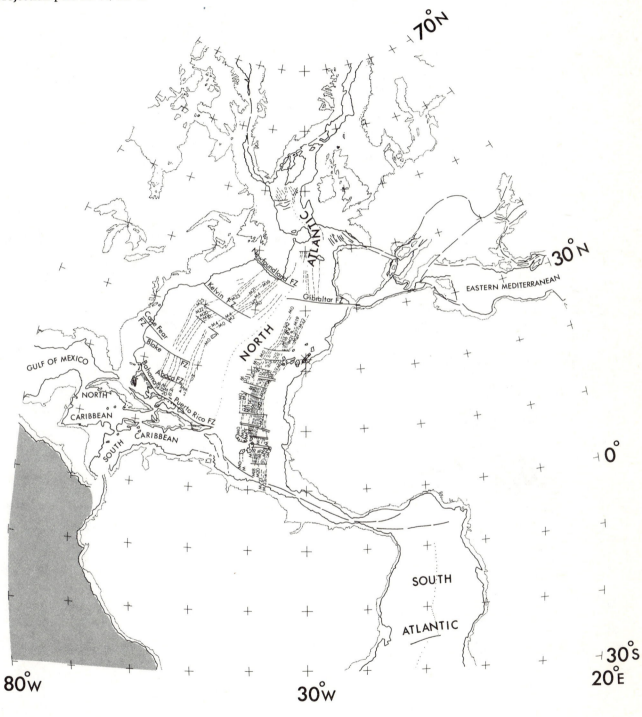

70°N

30°N

EASTERN MEDITERRANEAN

ATLANTIC

Newfoundland FZ

Kelvin FZ

Gibraltar FZ

Cape Fear FZ

Blake FZ

NORTH

GULF OF MEXICO

Bahama FZ

Atlaca FZ

NORTH CARIBBEAN

Puerto Rico FZ

SOUTH CARIBBEAN

0°

SOUTH

ATLANTIC

30°S
20°E

80°W

30°W

Map 19

Anomaly M7 (120 Ma) Hauterivian
modern dimensions Earth
Projection pole 22°N, 30°W

Map 20

Anomaly M7 (120 Ma) Hauterivian
ca 87% of modern diameter
Projection pole 22°N, 30°W

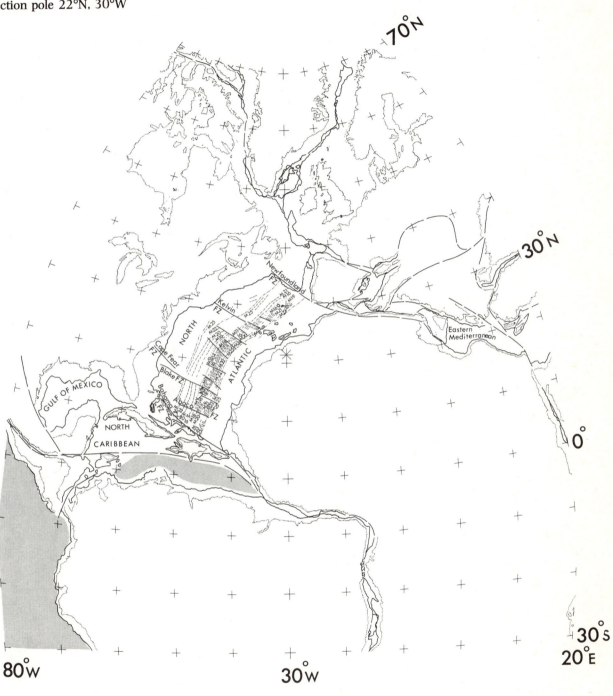

70°N

30°N

0°

30°S

20°E

80°W

30°W

Newfoundland FZ

Kelvin FZ

NORTH

Cape Fear FZ

Blake FZ

Bahama FZ

Atlantis FZ

ATLANTIC

GULF OF MEXICO

NORTH

CARIBBEAN

Eastern Mediterranean

Map 21

Anomaly M23 (146 Ma) Oxfordian
ca 84% of modern diameter
Projection pole 22°N, 30°W

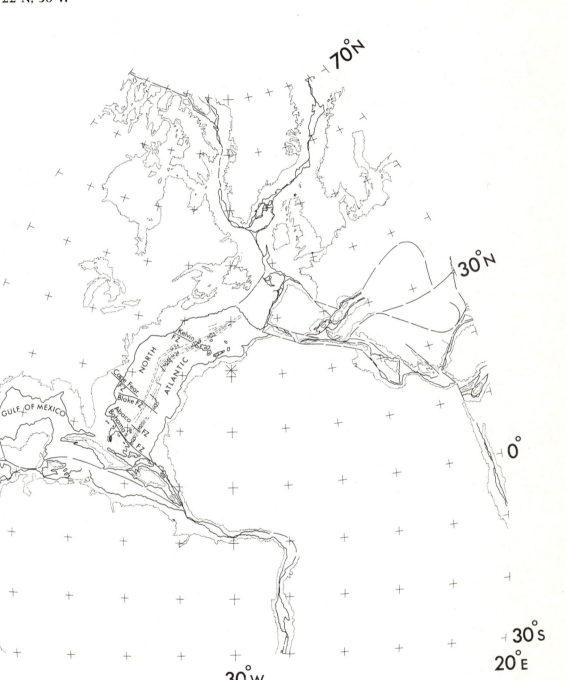

70°N

30°N

Kelvin FZ

NORTH ATLANTIC

Cape Fear FZ

Blake FZ

Abaco FZ

Bahama FZ

GULF OF MEXICO

0°

30°S
20°E

80°W

30°W

Map 22

Pangaea (180–200 Ma) late Triassic to lower Jurassic
modern dimensions Earth
Projection pole 22°N, 30°W

70°N

TETHYAN
OCEAN

30°N

30°N

0°

80°W

30°W

30°S

20°E

Map 23

Pangaea (180–200 Ma) late Triassic to lower Jurassic
80% of modern diameter
Projection pole 22°N, 30°W

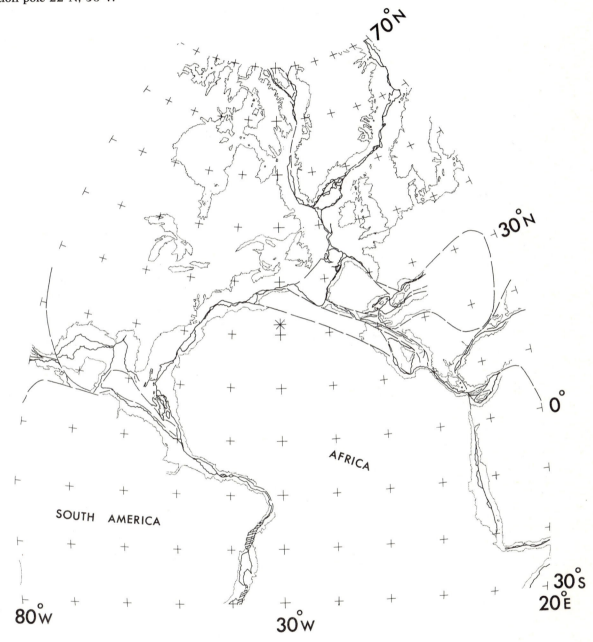

SECTION 3

South Atlantic

MAPS 24–33

(azimuthal equidistant projection; oblique case)

Map 24

Modern
For sources see text
Projection pole 22°S, 10°W

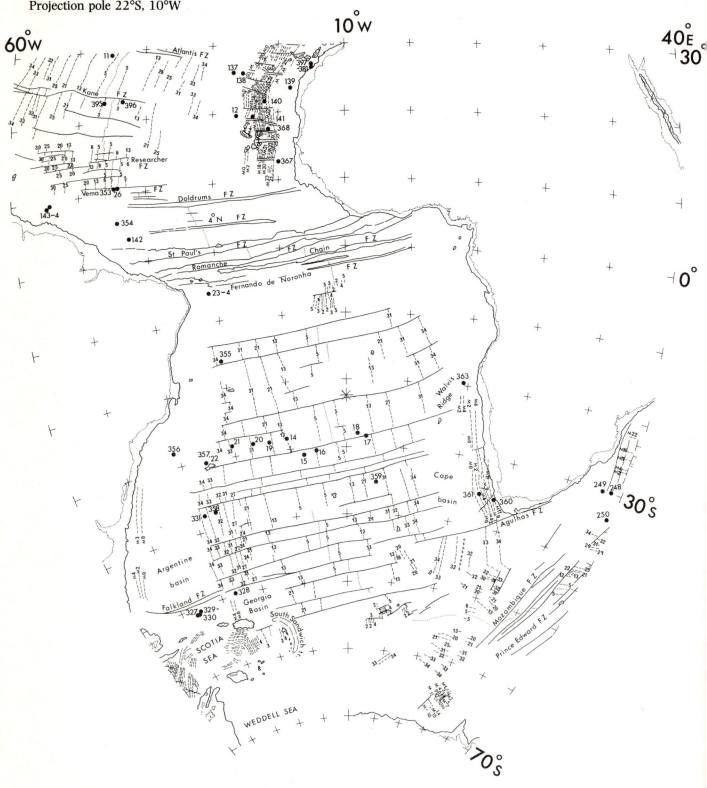

Map 25

Anomaly 9 (29 Ma) Oligocene
ca 97% of modern diameter
Projection pole 22°S, 10°W

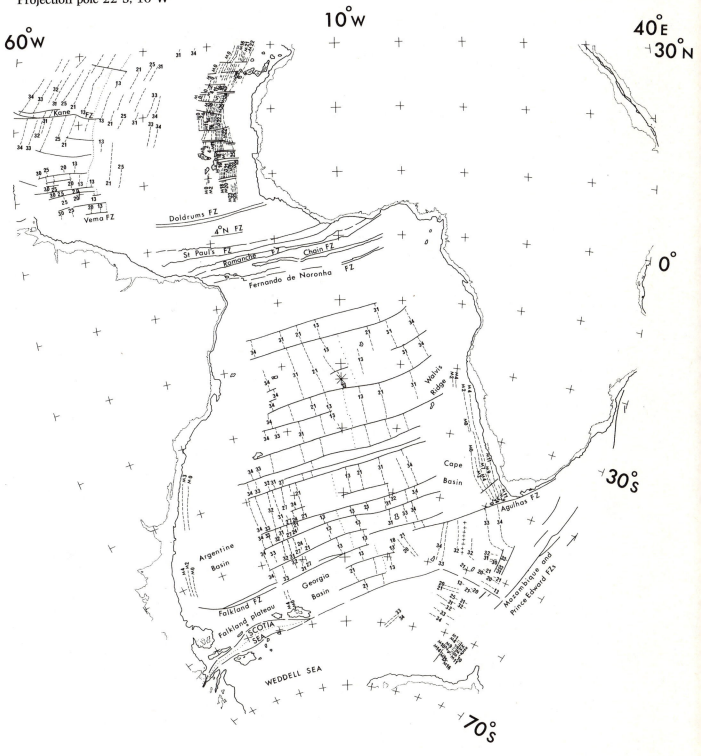

60°W

10°W

40°E
30°N

Kane FZ

31 34

25 31
13
34 33
32
31 25 21
13
34 33
21 25
31 33 34
34 32
13
34 34
21
13

30 25 20 13
30 25 20 13 25
25 20 13
30 25 20 13
25 20 13
30 25 20 13
Vema FZ

Doldrums FZ

4°N FZ

St Paul's FZ
Romanche FZ   Chain FZ
Fernando de Noronha FZ

0°

31
13 21 31 34
34
21 31
34
31 21
34 13 21
34 33 31
Walvis Ridge

34 33
32 31 21
34 33
32 27 21
34 33
32 27 21 31 34
32
34 31
27 21
34 33 21 24
34 33 21 24
34 21 24 31 27
34
34

Cape Basin

Agulhas FZ

33 34
32 32 32
31 30

34
33
18 13
21 13
20
21 21 20 13
25 21
25 31
32 33

Argentine Basin

Georgia Basin

Falkland FZ
Falkland plateau
SCOTIA SEA

33 34
34

Mozambique and Prince Edward FZs

30°S

WEDDELL SEA

70°S

Map 26

Anomaly 24 (56 Ma) Palaeocene
modern dimensions Earth
Projection pole 22°S, 20°W

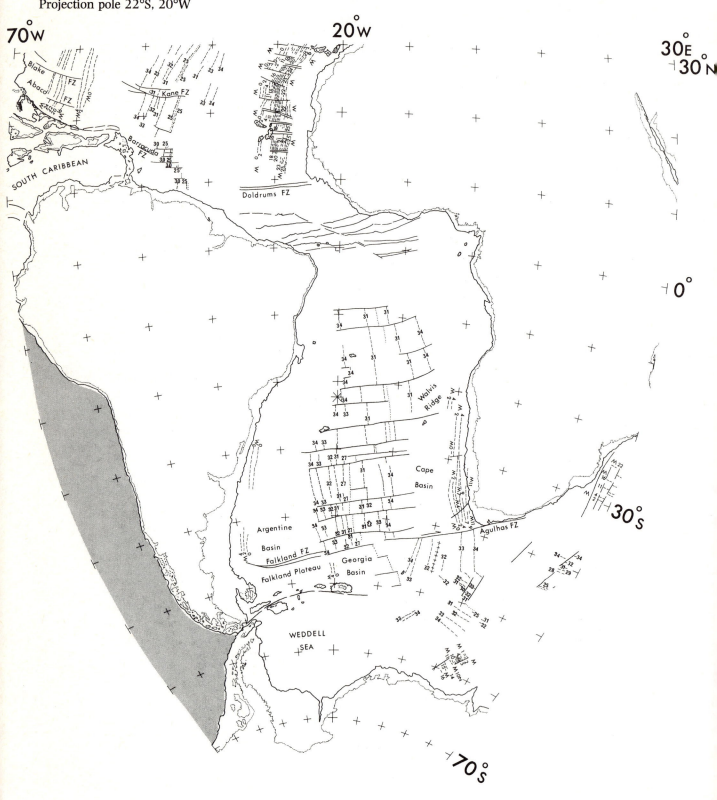

Map 27

Anomaly 24 (56 Ma) Palaeocene
ca 94% of modern diameter
Projection pole 22°S, 30°W

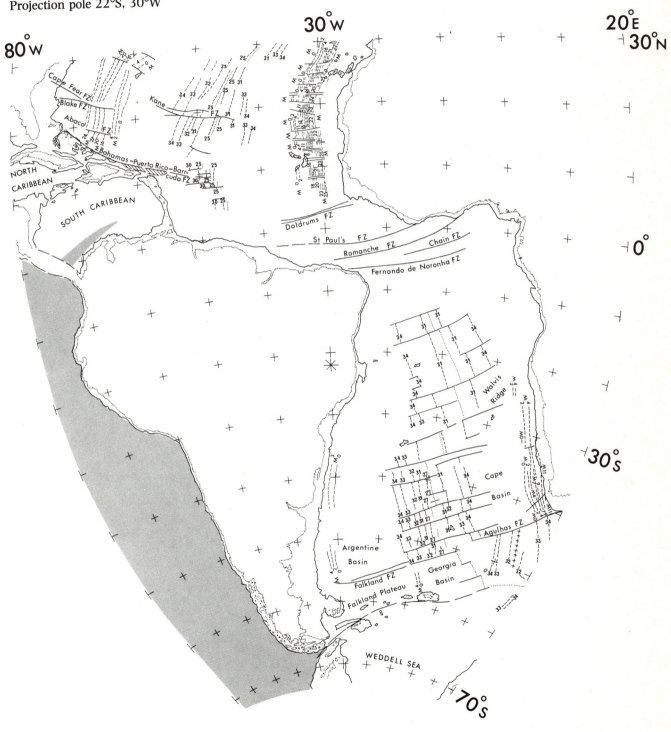

79

Map 28

Turonian (90 Ma) upper Cretaceous
90% of modern diameter
Projection pole 22°S, 20°W

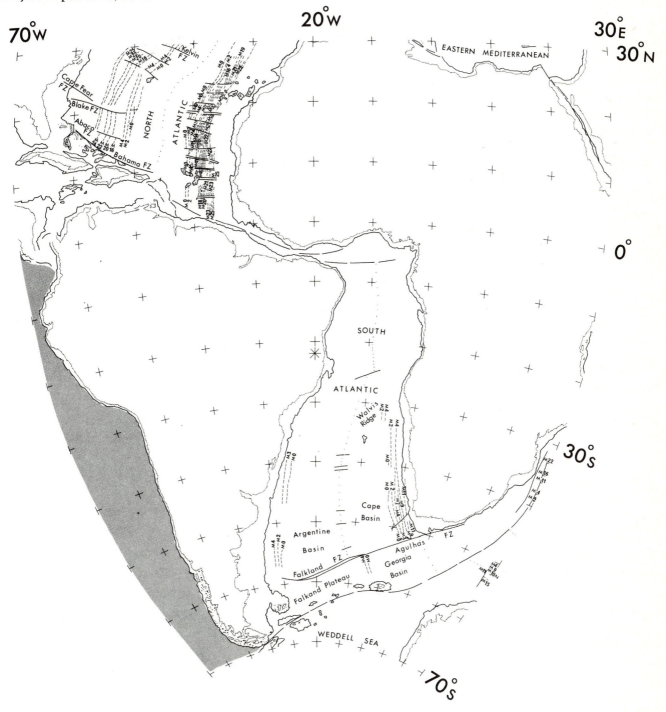

Map 29

Anomaly M7 (120 Ma) Hauterivian
modern dimensions Earth
Projection pole 22°S, 10°W

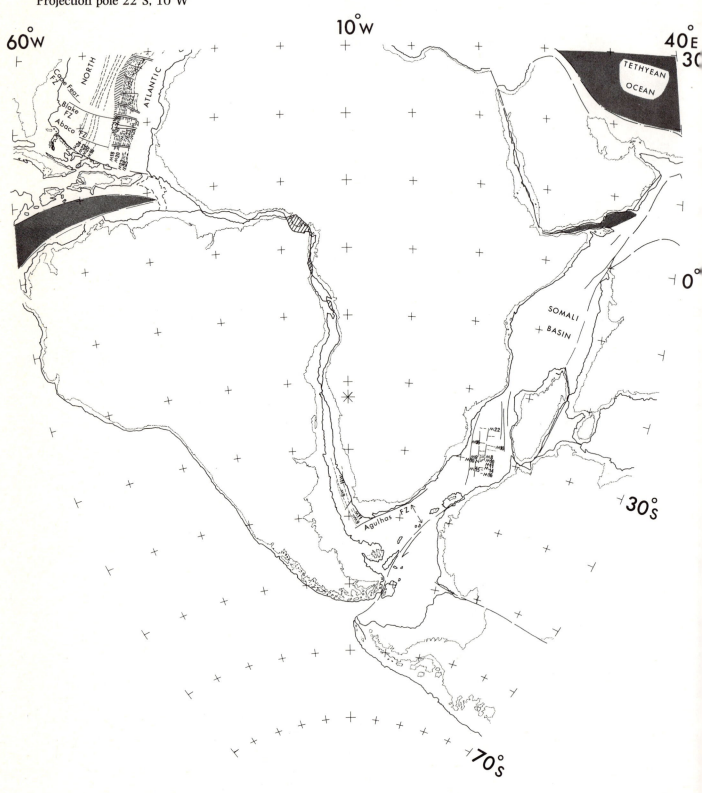

Map 30

Anomaly M7 (120 Ma) Hauterivian
ca 87% of modern diameter
Projection pole 22°S, 20°W

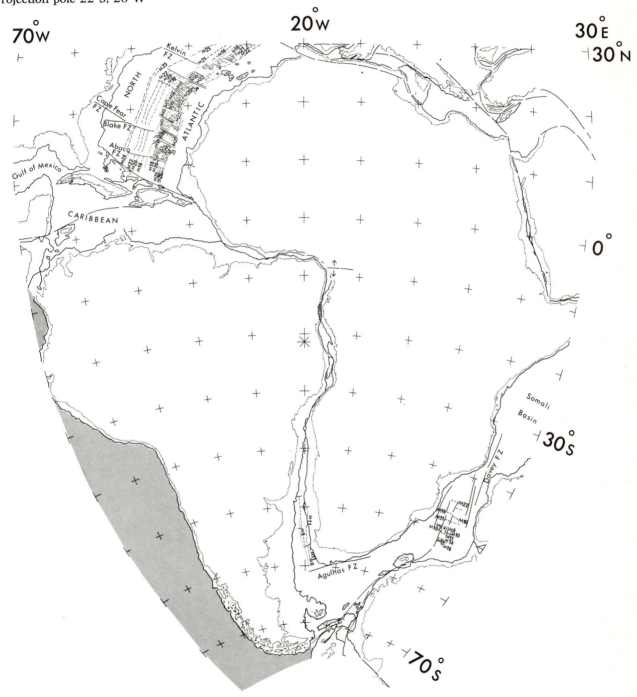

Map 31

Anomaly M23 (146 Ma) Oxfordian
ca 84% of modern diameter
Projection pole 22°S, 30°W

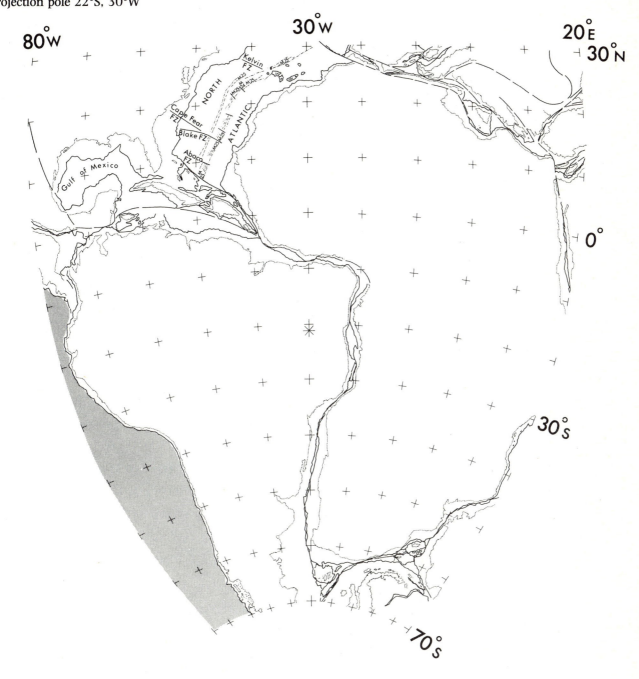

Map 32

Pangaea (180–200 Ma) late Triassic to lower Jurassic
modern dimensions Earth
Projection pole 22°S, 20°W

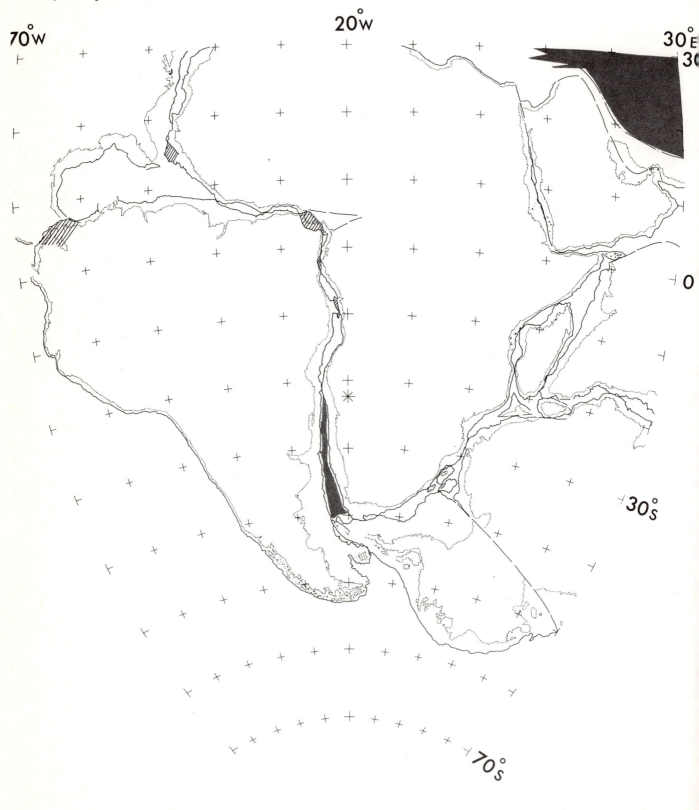

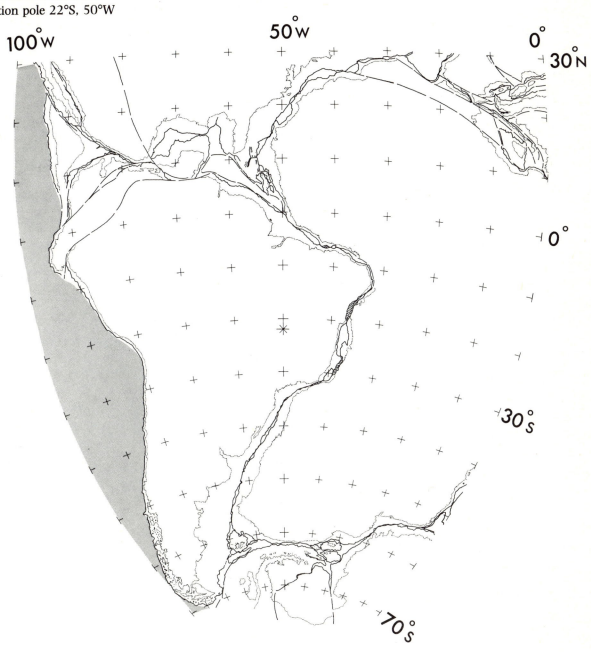

Map 33

Pangaea (180–200 Ma) late Triassic to lower Jurassic
80% of modern diameter
Projection pole 22°S, 50°W

100°W
50°W
0°
30°N
0°
30°S
70°S

SECTION 4

Indian Ocean

MAPS 34–43

(azimuthal equidistant projection; oblique case)

Map 34

Modern
For sources see text
Projection pole 22°S, 80°E

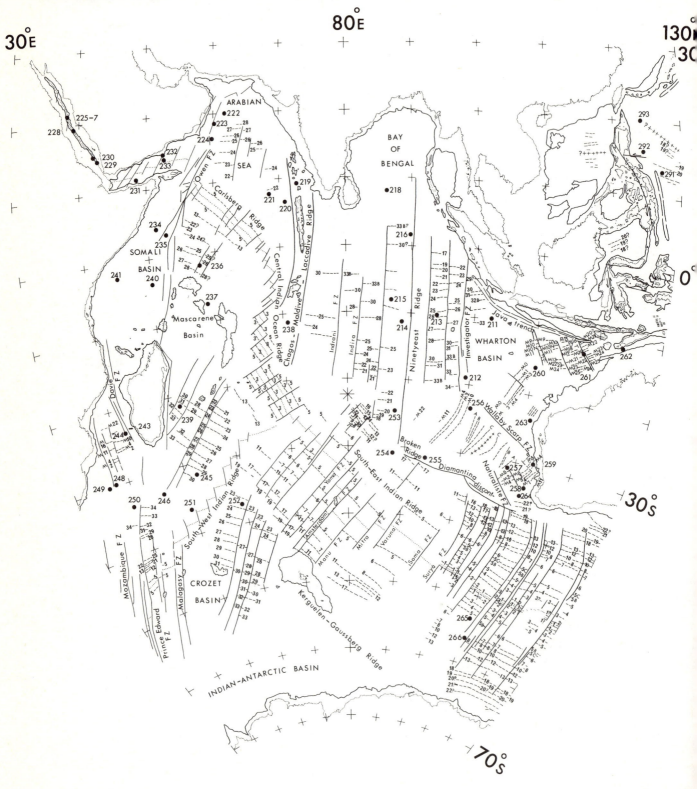

Map 35

Anomaly 9 (29 Ma) Oligocene
ca 97% of modern diameter
Projection pole 22°S, 80°E

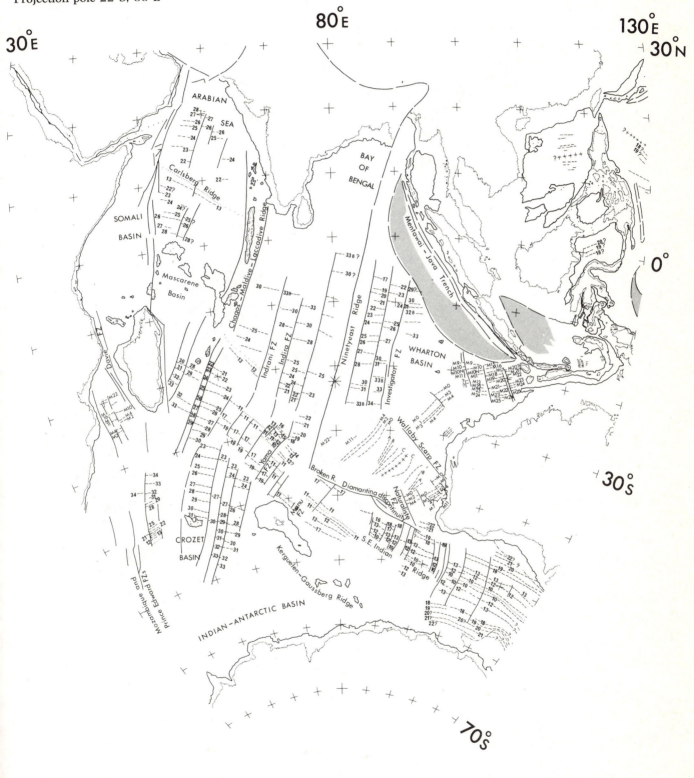

Map 36

Anomaly 24 (56 Ma) Palaeocene
modern dimensions Earth
Projection pole 22°S, 70°E

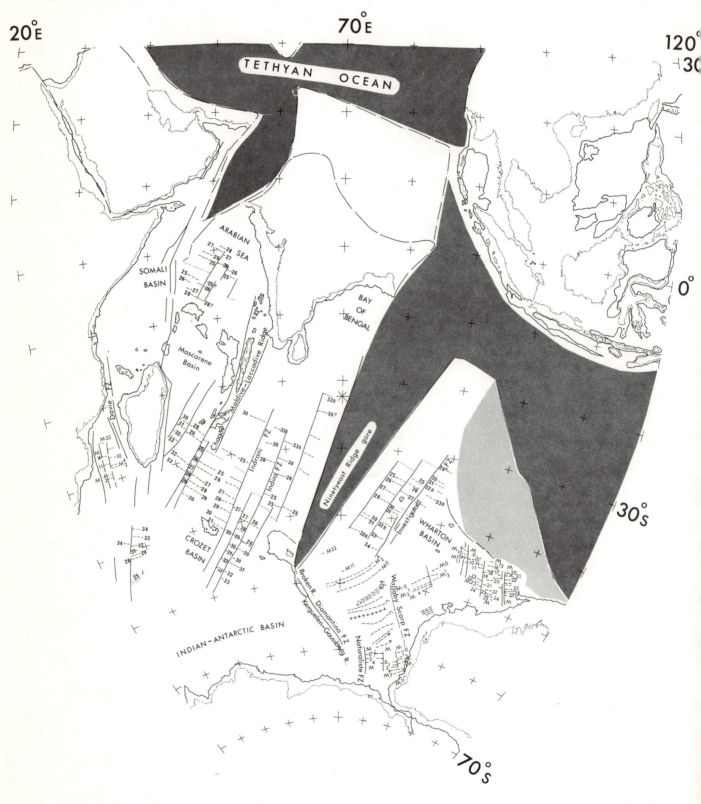

Map 37

Anomaly 24 (56 Ma) Palaeocene
ca 94% of modern diameter
Projection pole 22°S, 60°E

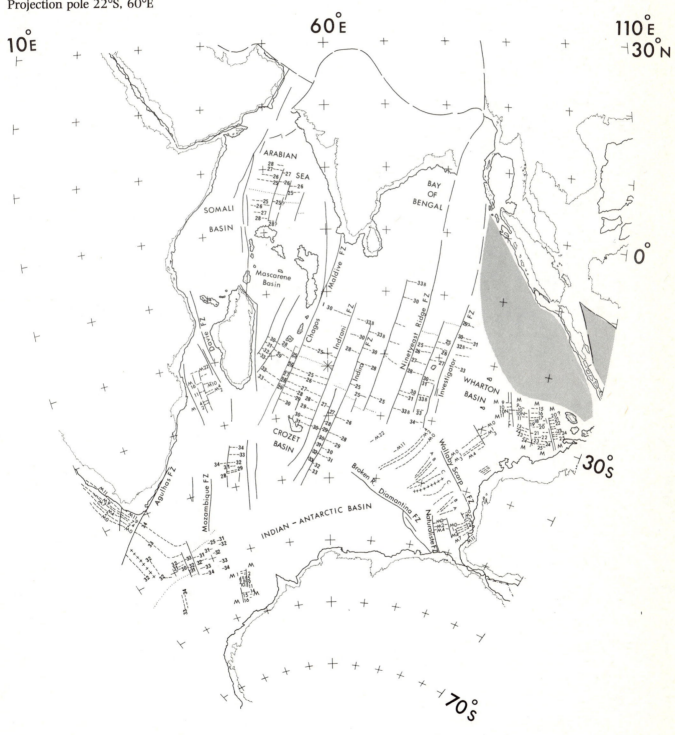

Map 38

Turonian (90 Ma) upper Cretaceous
90% of modern diameter
Projection pole 22°S, 70°E

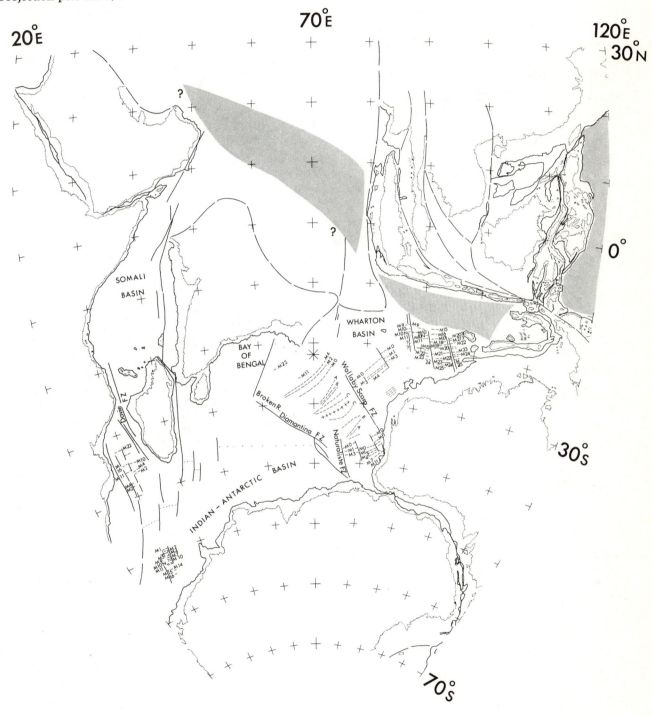

20°E

70°E

120°E
30°N

?

?

0°

SOMALI
BASIN

WHARTON
BASIN

M9
M10
M11

M9
M10
M11

M10
M11

M15

M23
M21

M20
M21
M22

BAY
OF
BENGAL

M22

M24

M1
M0

M0
M3
M4

M22
M21
M25
24

M20
M24

M22
M23
M24
M25

Wallaby Scarp F.Z.

M11

M0
M3
M4

Broken R. Diamantina F.Z.

M0

Davie F.Z.

M22

Naturaliste F.Z.

M0
M3
M4
M11

M10
M4
M2

30°S

INDIAN – ANTARCTIC   BASIN

M1
M5
M10N
M11
M15
M10

M2
M10
M14

M14

70°S

Map **39**

Anomaly M7 (120 Ma) Hauterivian
modern dimensions Earth
Projection pole 22°S, 80°E

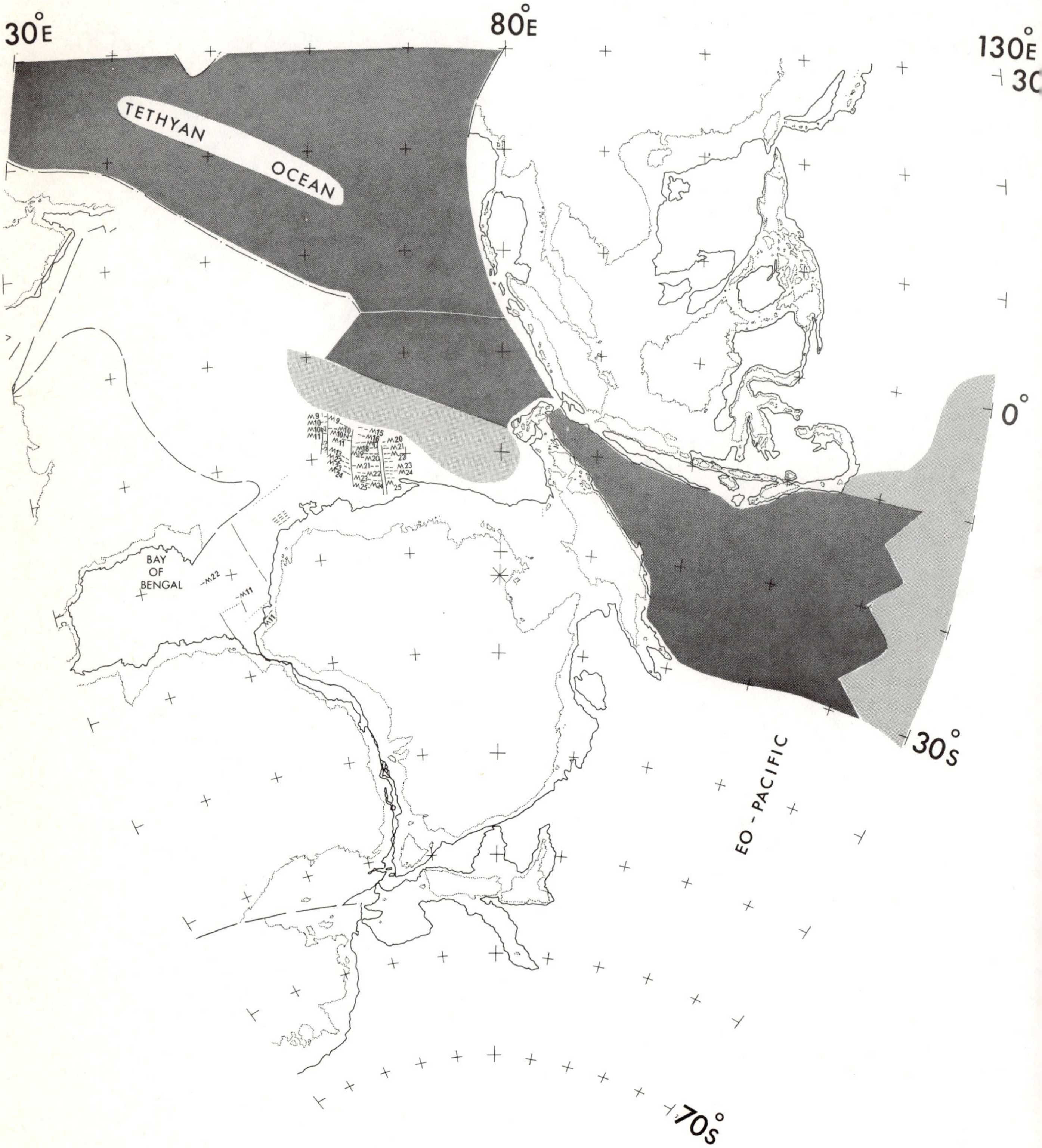

Map 40

Anomaly M7 (120 Ma) Hauterivian
ca 87% of modern diameter
Projection pole 22°S, 70°E

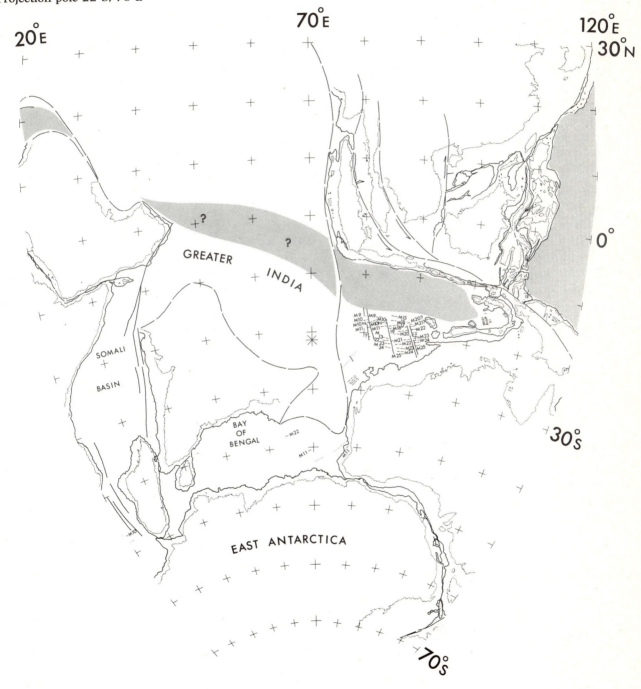

20°E

70°E

120°E
30°N

0°

GREATER

?

?

INDIA

SOMALI
BASIN

BAY
OF
BENGAL

M9
M10
M10
M11

M10
M10
M18
M

M15
M19
M20

M20?
M21?
M22

30°S

M22
M23
M24

M21
M22
M23
M24

M22
M23
M24
M25

M25

M25

EAST ANTARCTICA

70°S

Map 41

Anomaly M23 (146 Ma) Oxfordian
ca 84% of modern diameter
Projection pole 22°S, 60°E

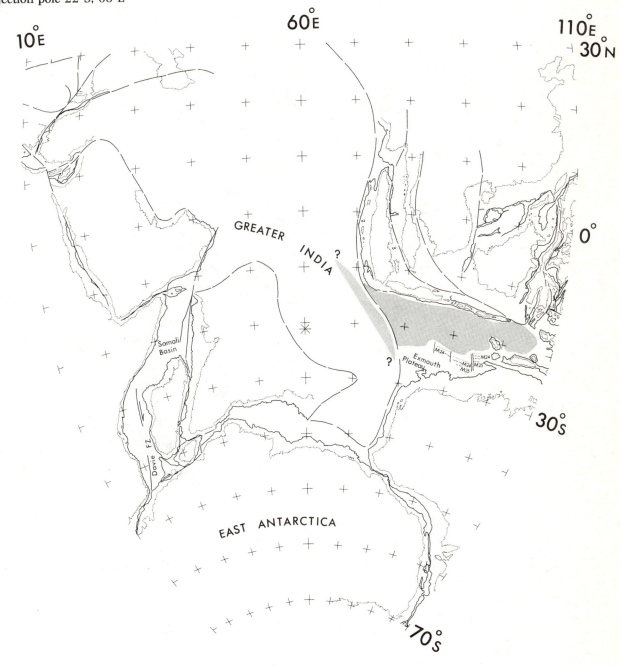

10°E

60°E

110°E
30°N

GREATER INDIA ?

0°

?

Somali Basin

Exmouth Plateau

M24

M24
M25

M24
M25

Davie FZ

30°S

EAST ANTARCTICA

70°S

Map 42

Pangaea (180–200Ma) late Triassic to lower Jurassic
modern dimensions Earth
Projection pole 22°S, 70°E

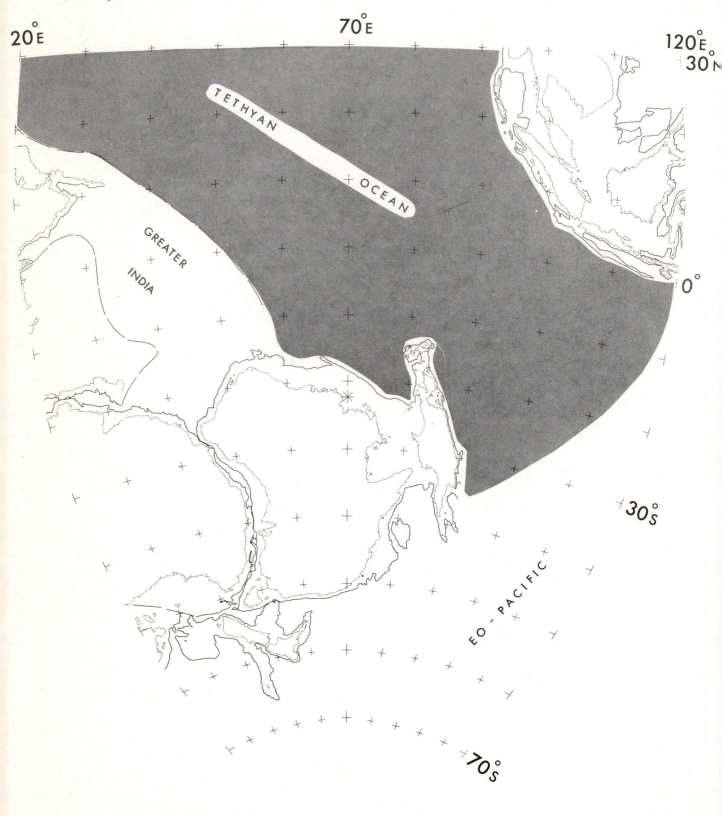

Map 43

Pangaea (180–200 Ma) late Triassic to lower Jurassic
80% of modern diameter
Projection pole 22°S, 40°E

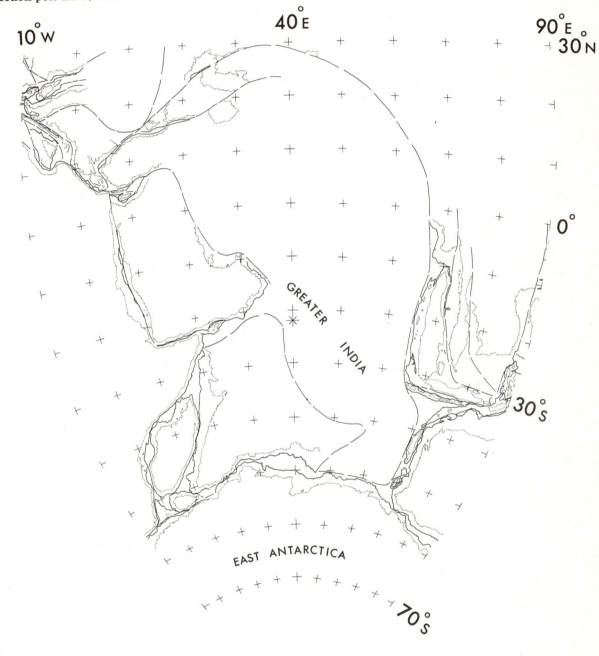

SECTION 5
North and Central Pacific
MAPS 44–53
(azimuthal equidistant projection; modified equatorial case)

Map 44

Modern
For sources see text
Projection pole 0°, 170°W

Map 45

Anomaly 9 (29 Ma) Oligocene
ca 97% of modern diameter
Projection pole 0°, 170°W

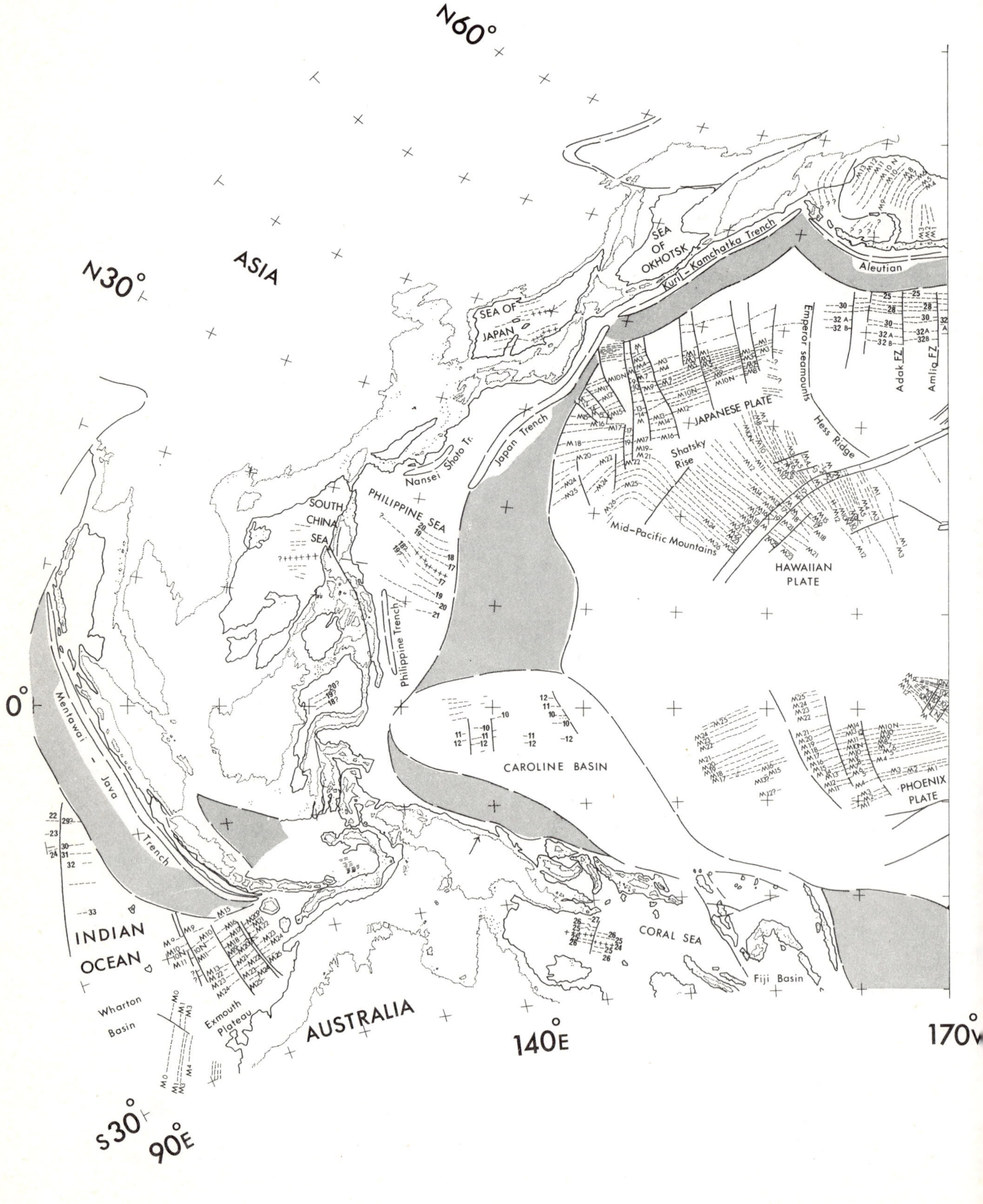

N60°

N30°

ASIA

SEA OF OKHOTSK

Kuril - Kamchatka Trench

Aleutian

SEA OF JAPAN

Emperor seamounts

Adak FZ

Amlia FZ

Nansei Shoto Tr.

Japan Trench

JAPANESE PLATE

Hess Ridge

Shatsky Rise

PHILIPPINE SEA

SOUTH CHINA SEA

Mid-Pacific Mountains

HAWAIIAN PLATE

Philippine Trench

Mentawai - Java

0°

CAROLINE BASIN

PHOENIX PLATE

Trench

22
23
24

29
30
31
32

33

INDIAN OCEAN

CORAL SEA

Fiji Basin

Wharton Basin

Exmouth Plateau

AUSTRALIA

140°E

170°W

S30°

90°E

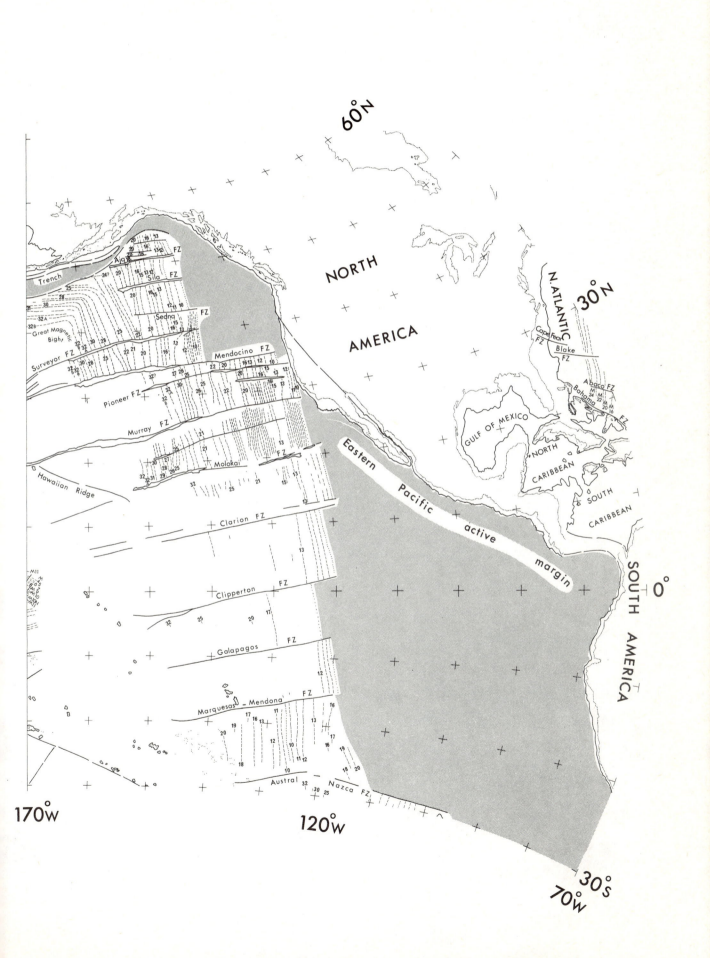

Map 46

Anomaly 24 (56 Ma) Palaeocene
modern dimensions Earth
Projection pole 0°, 150°W

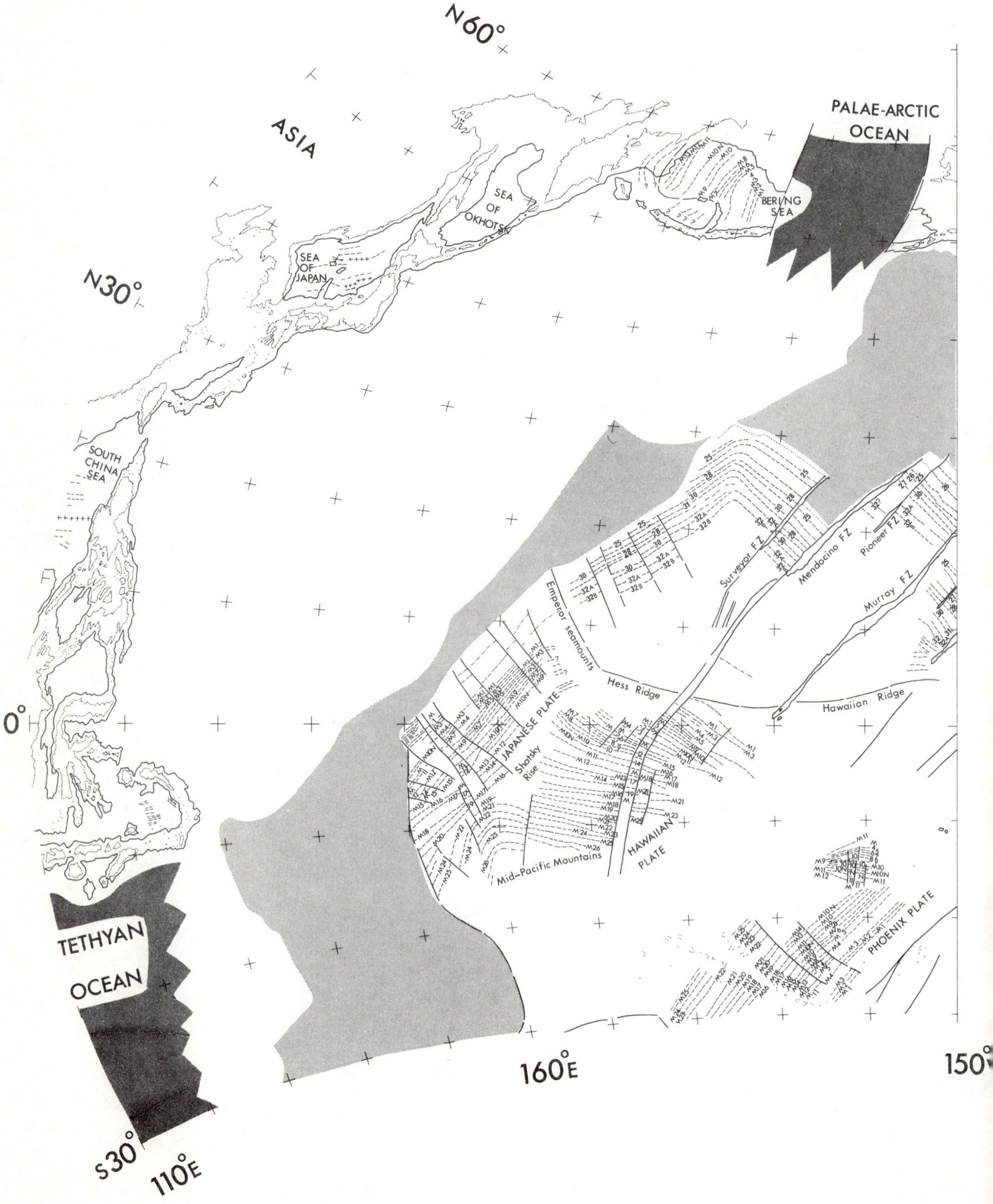

N60°

ASIA

PALAE-ARCTIC
OCEAN

SEA
OF
OKHOTSK

BERING
SEA

N30°

SEA
OF
JAPAN

SOUTH
CHINA
SEA

Surveyor F.Z.

Mendocino FZ

Pioneer FZ

Murray FZ

Emperor seamounts

Hess Ridge

Hawaiian Ridge

JAPANESE PLATE

Shatsky
Rise

HAWAIIAN
PLATE

Mid-Pacific Mountains

PHOENIX PLATE

0°

TETHYAN

OCEAN

S30°

110°E

160°E

150°

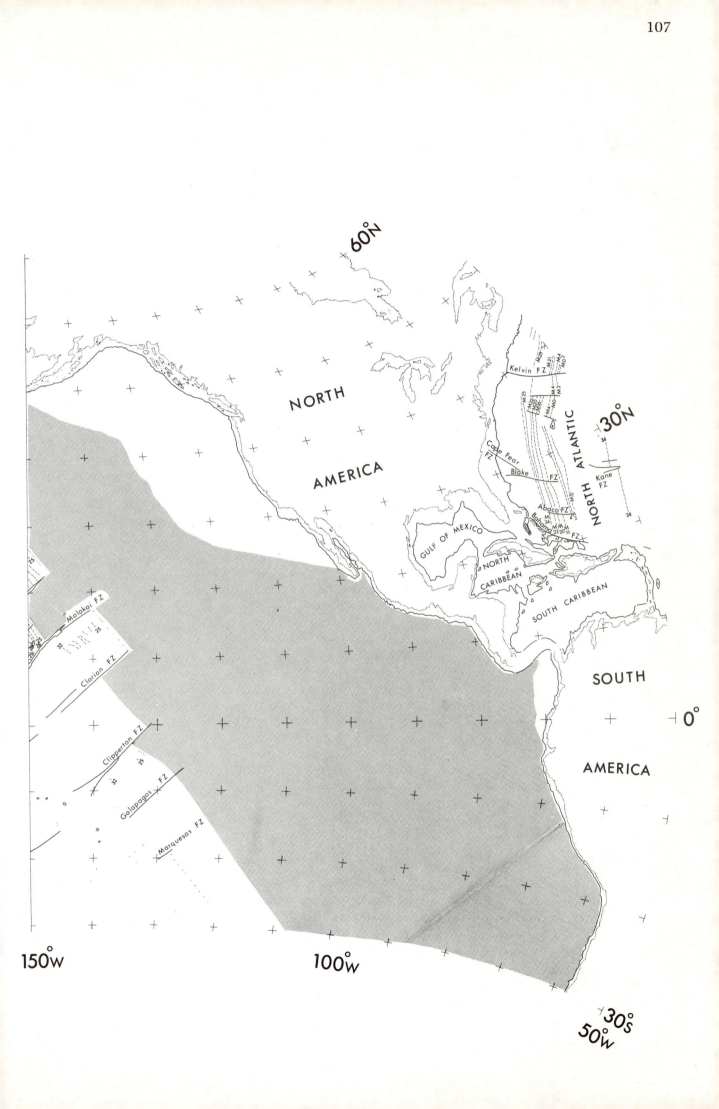

Map 47

Anomaly 24 (56 Ma) Palaeocene
ca 94% of modern diameter
Projection pole 0°, 160°W

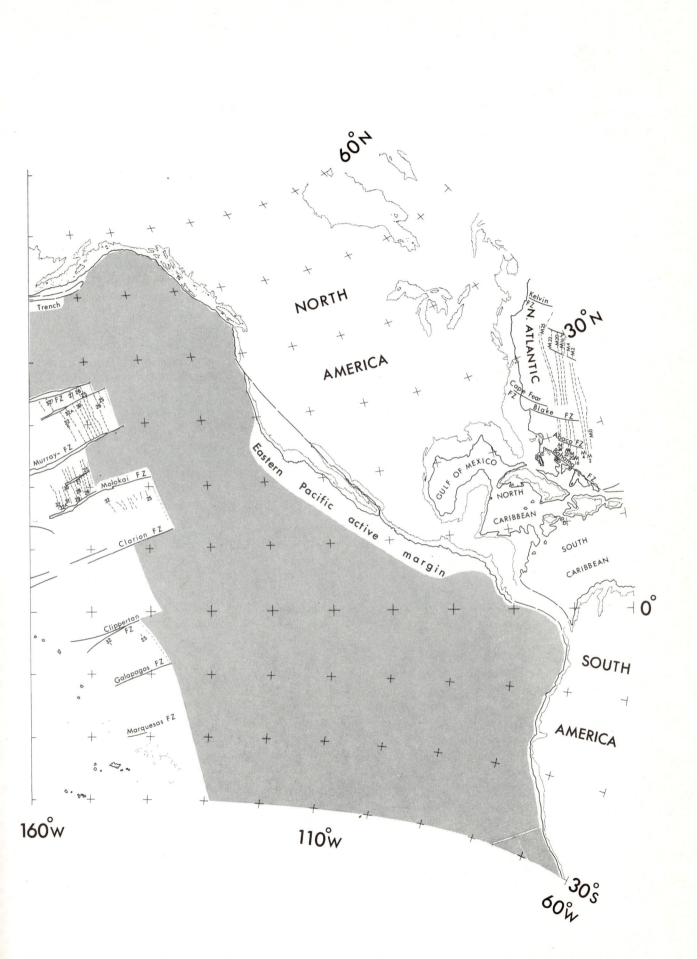

Map 48

Turonian (90 Ma) upper Cretaceous
90% of modern diameter
Projection pole 0°, 150°W

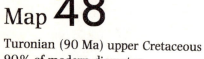

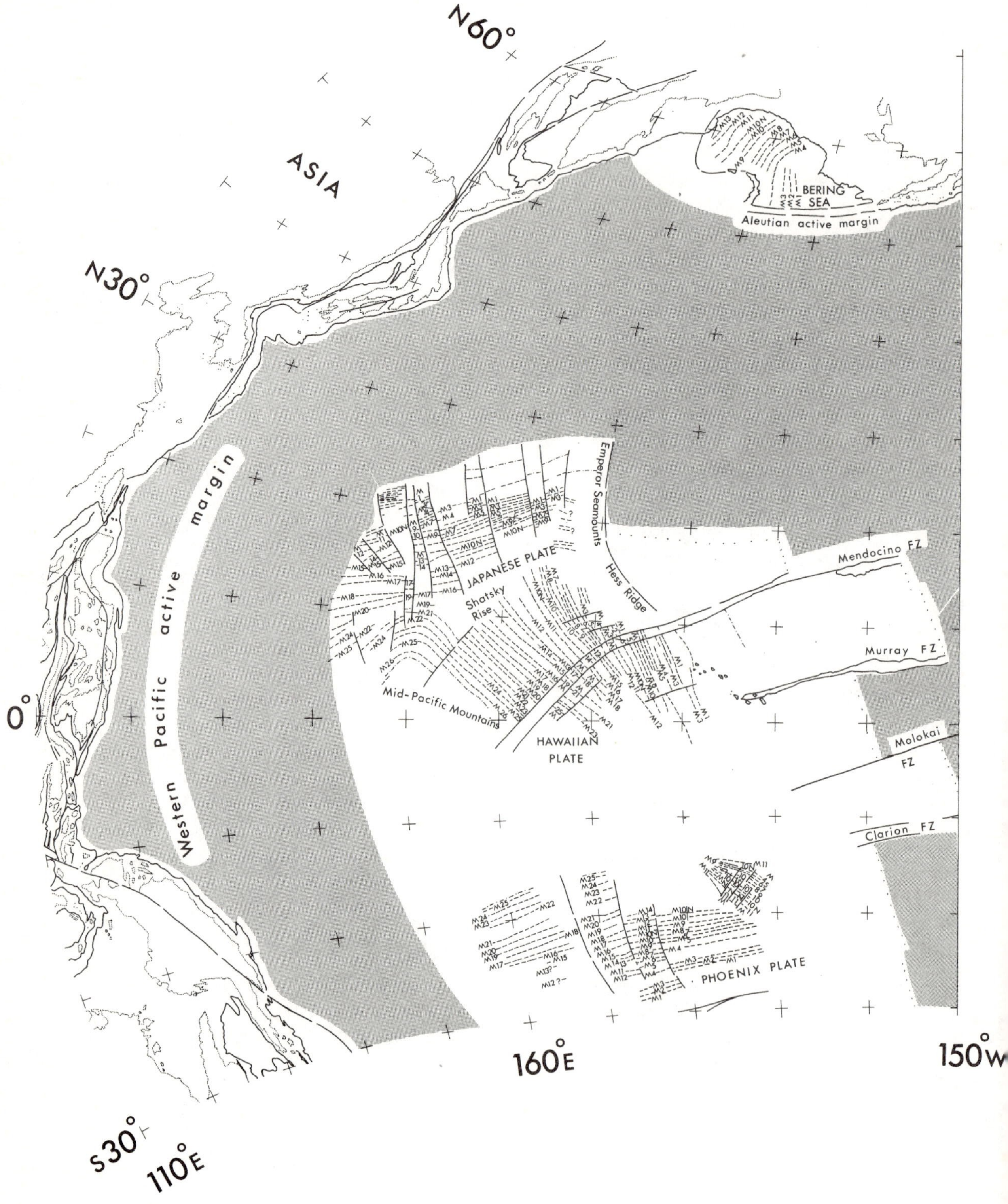

N60°

ASIA

N30°

BERING
SEA

Aleutian active margin

Western Pacific active margin

Emperor Seamounts

JAPANESE PLATE

Hess Ridge

Shatsky
Rise

Mendocino FZ

Murray FZ

Mid-Pacific Mountains

HAWAIIAN
PLATE

Molokai
FZ

Clarion FZ

0°

PHOENIX PLATE

160°E

150°W

S30°

110°E

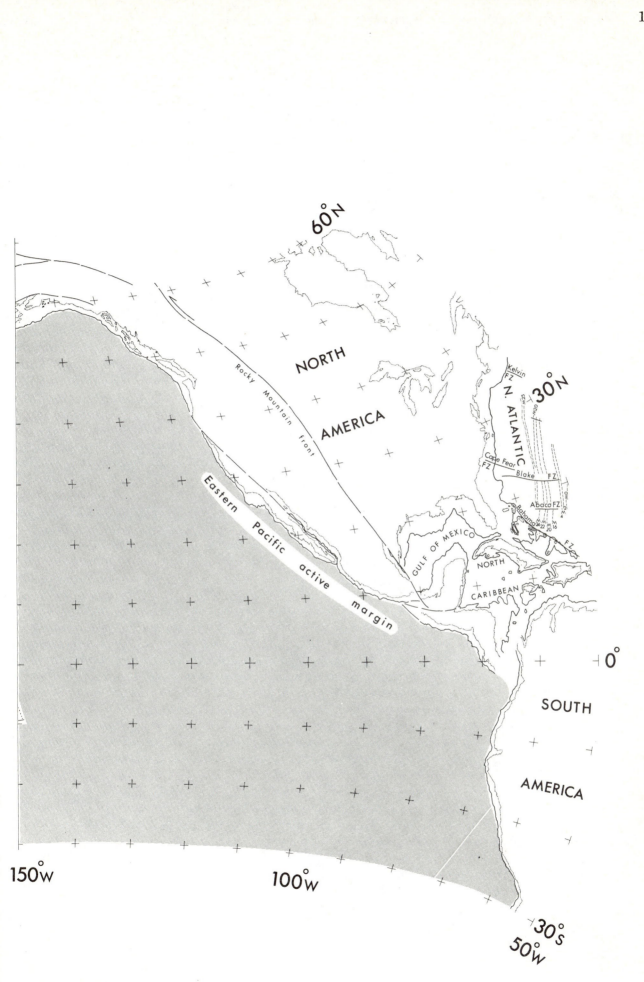

Map **49**

Anomaly M7 (120 Ma) Hauterivian
modern dimensions Earth
Projection pole 0°, 150°W

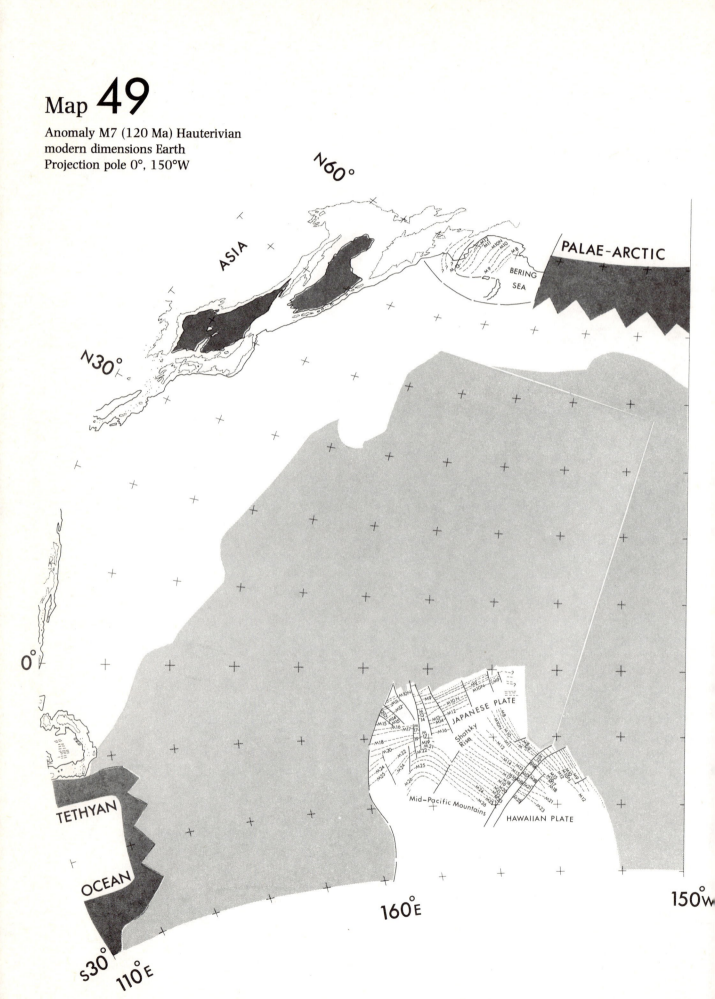

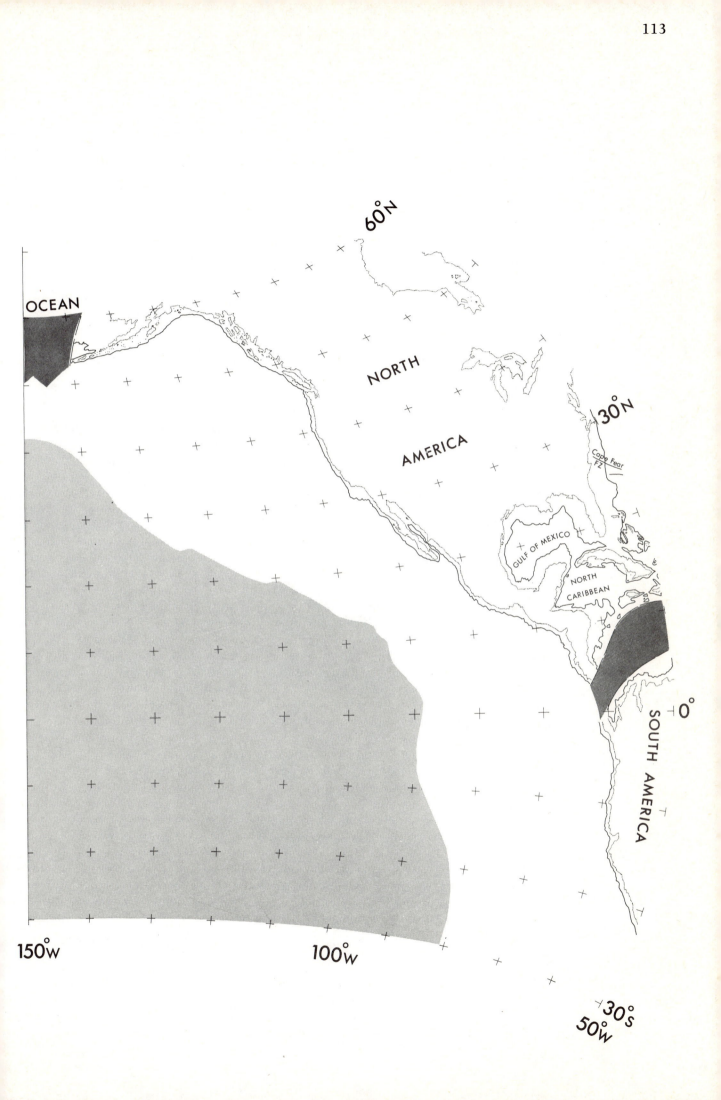

OCEAN

60°N

NORTH

AMERICA

30°N

Cape Fear
FZ

GULF OF MEXICO

NORTH
CARIBBEAN

0°

SOUTH AMERICA

30°S
50°W

150°W

100°W

Map 50

Anomaly M7 (120 Ma) Hauterivian
ca 87% of modern diameter
Projection pole 0°, 160°W

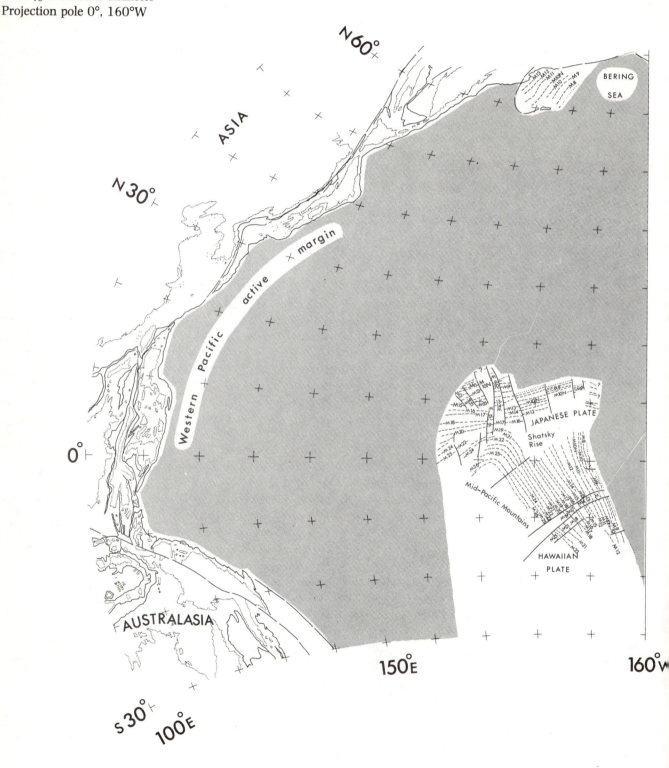

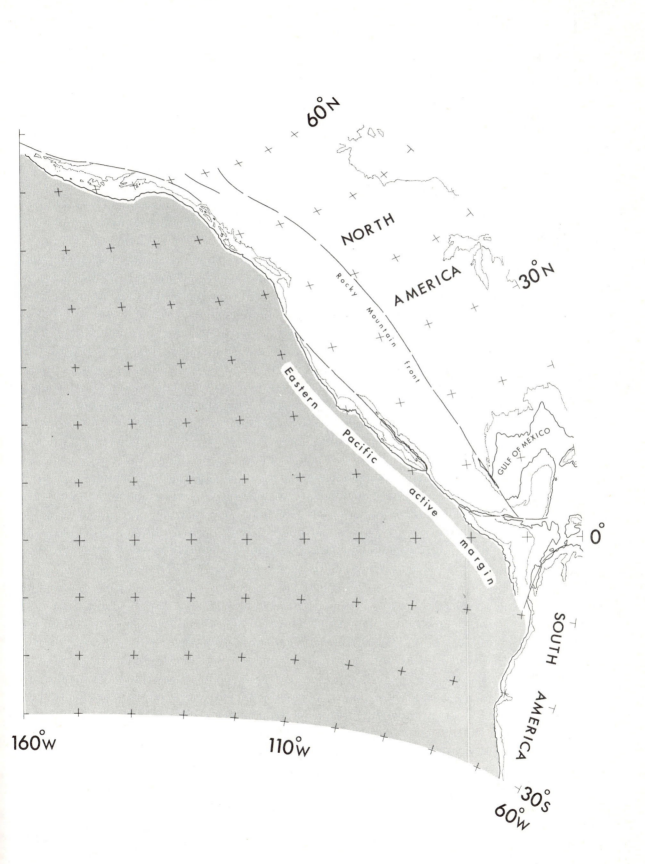

Map 51

Anomaly M23 (146 Ma) Oxfordian
ca 84% of modern diameter
Projection pole 0°, 170°W

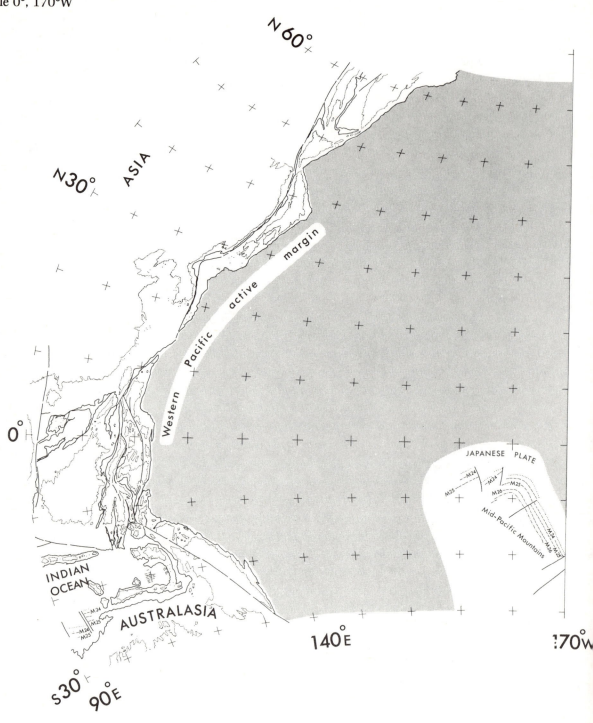

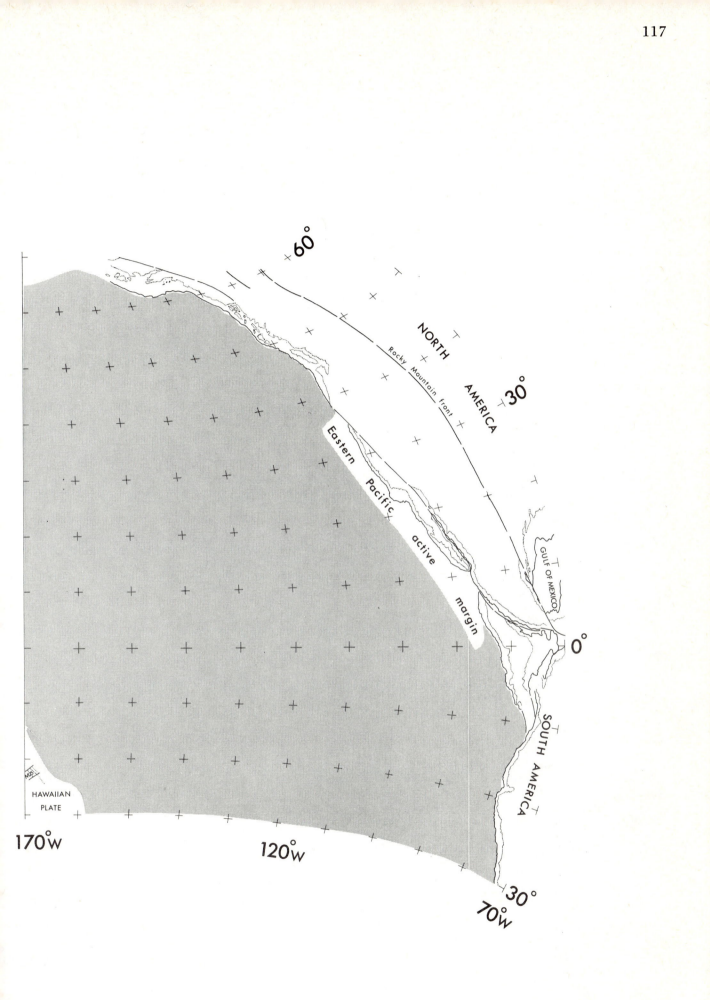

Map 52

Pangaea (180–200 Ma) late Triassic to lower Jurassic
modern dimensions Earth
Projection pole 0°, 160°W

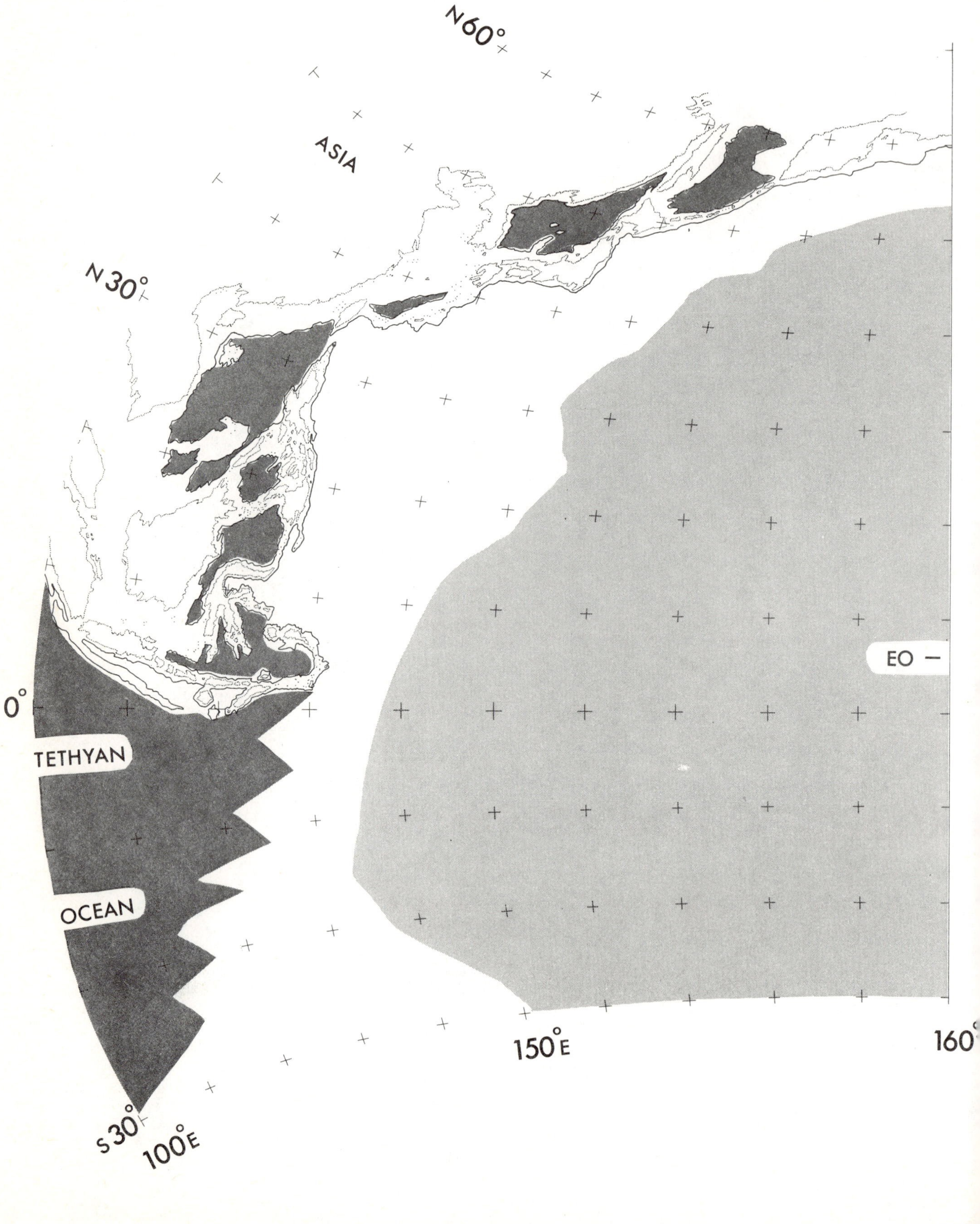

N 60°

ASIA

N 30°

0°

TETHYAN

OCEAN

EO —

150°E

160°

S 30°

100°E

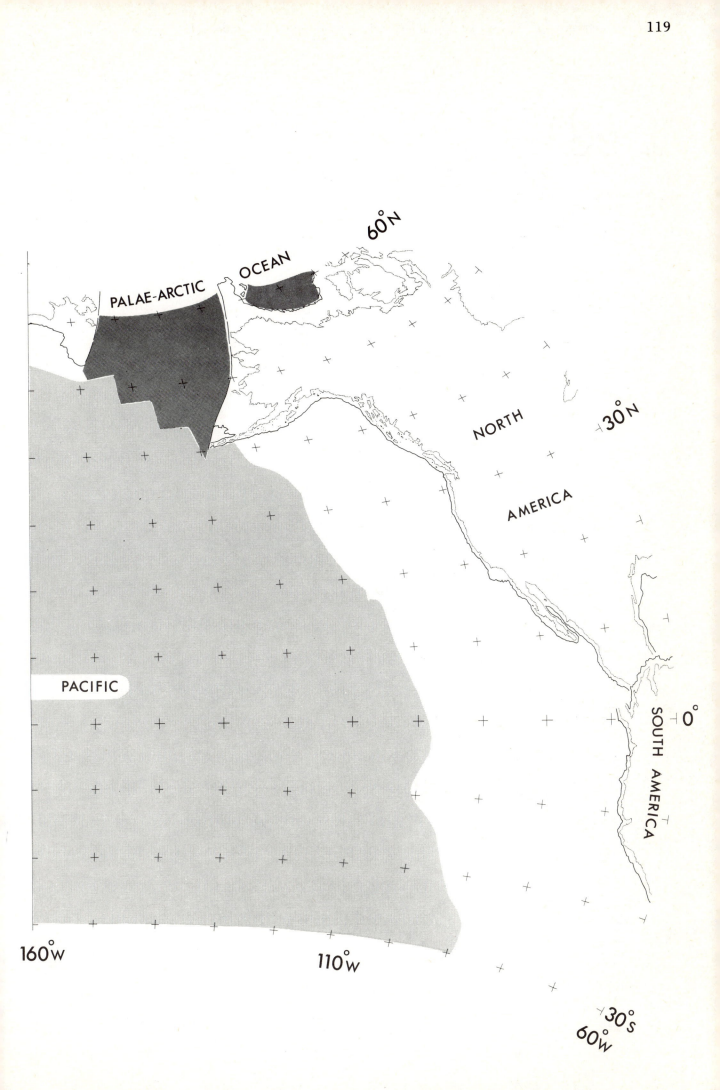

Map 53

Pangaea (180–200 Ma) late Triassic to lower Jurassic
80% of modern diameter
Projection pole 0°, 170°W

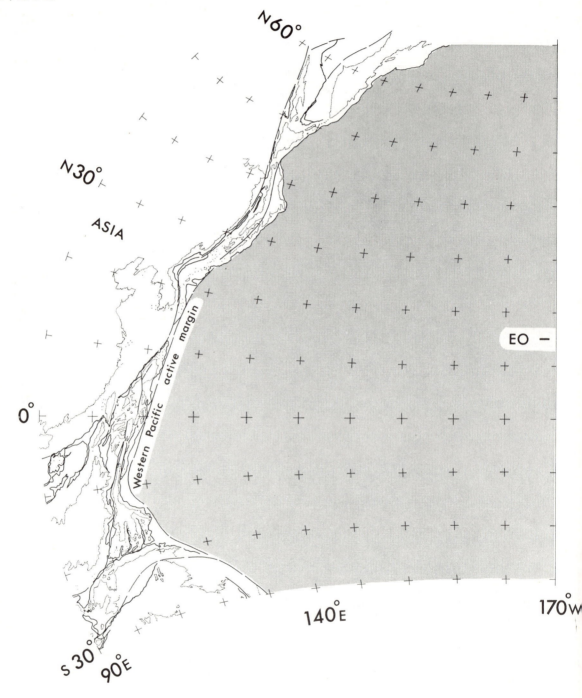

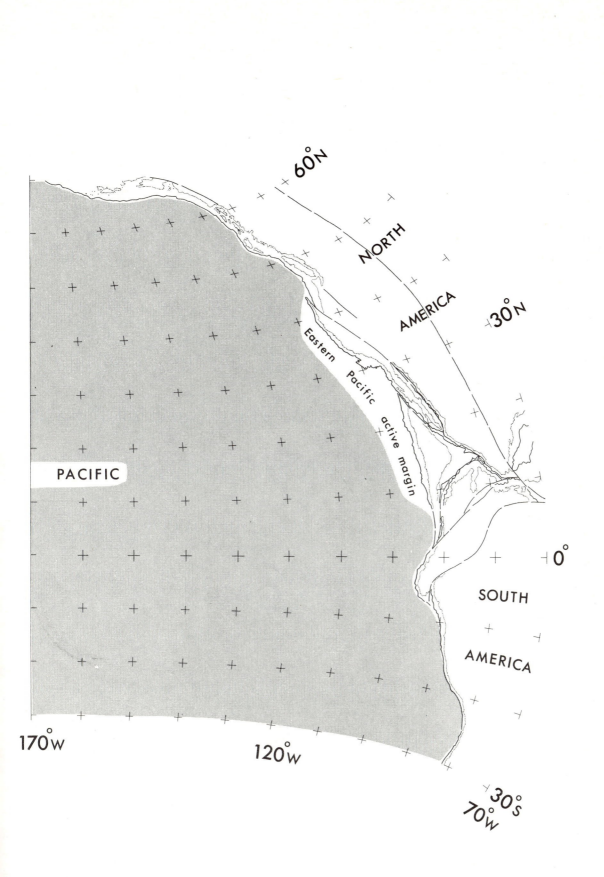

SECTION 6

Southern hemisphere north to 20° south latitude

MAPS 54–63

(azimuthal equidistant projection; polar case)

0°

Map 54

Modern
For sources see text

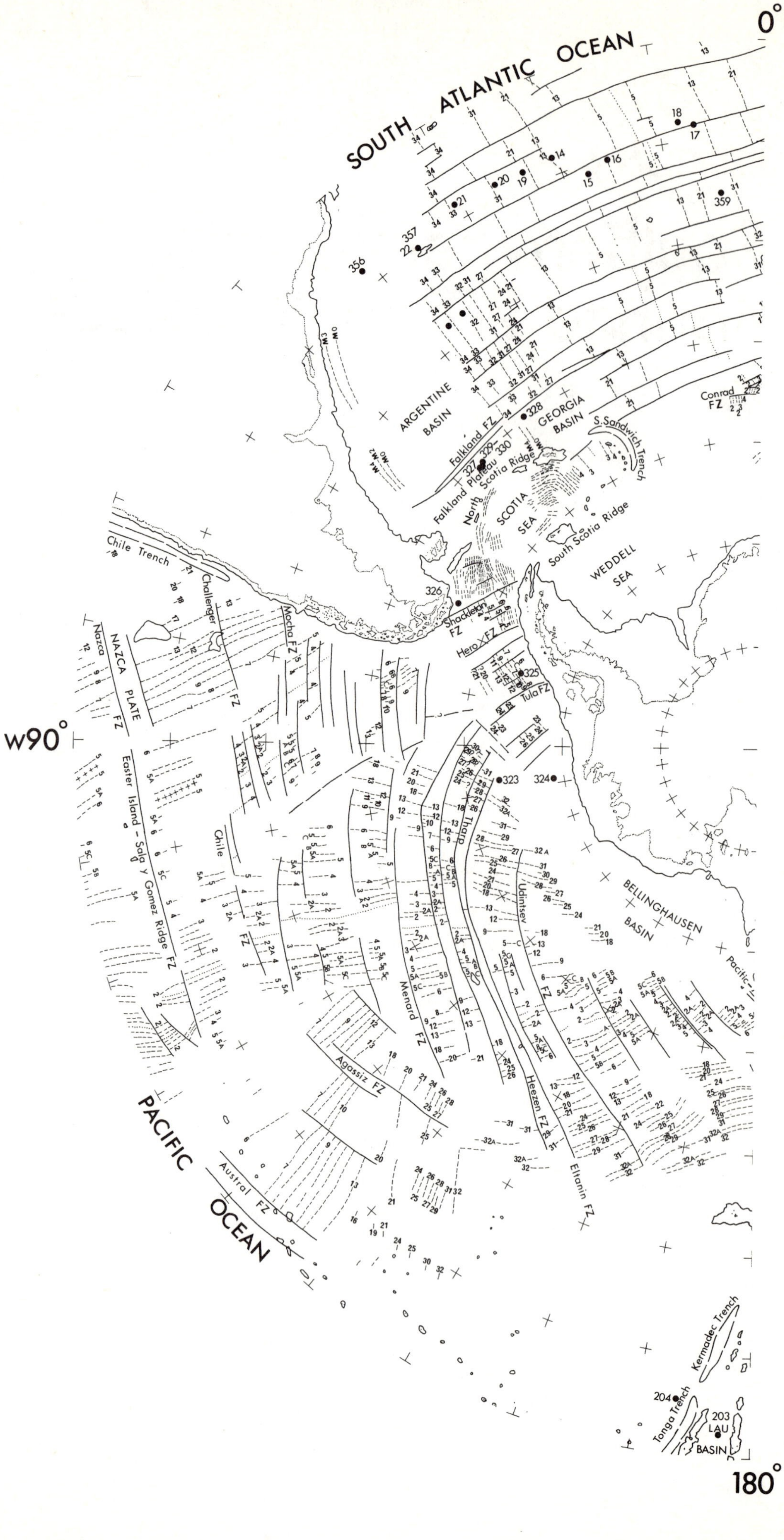

SOUTH ATLANTIC OCEAN

ARGENTINE BASIN

GEORGIA BASIN

Falkland FZ

Falkland Plateau

North Scotia Ridge

S. Sandwich Trench

Conrad FZ

SCOTIA SEA

WEDDELL SEA

South Scotia Ridge

Chile Trench

Challenger FZ

Mocha FZ

Shackleton FZ

Hero X FZ

Tula FZ

NAZCA PLATE

Nazca FZ

W90°

Easter Island – Sala y Gomez Ridge FZ

Chile FZ

Tharp FZ

Udintsev FZ

BELLINGHAUSEN BASIN

Pacific

Menard FZ

Agassiz FZ

Heezen FZ

Eltanin FZ

Austral FZ

PACIFIC OCEAN

Kermadec Trench

Tonga Trench

204

203 LAU BASIN

180°

Map 55

Anomaly 9 (29 Ma) Oligocene
ca 97% of modern diameter

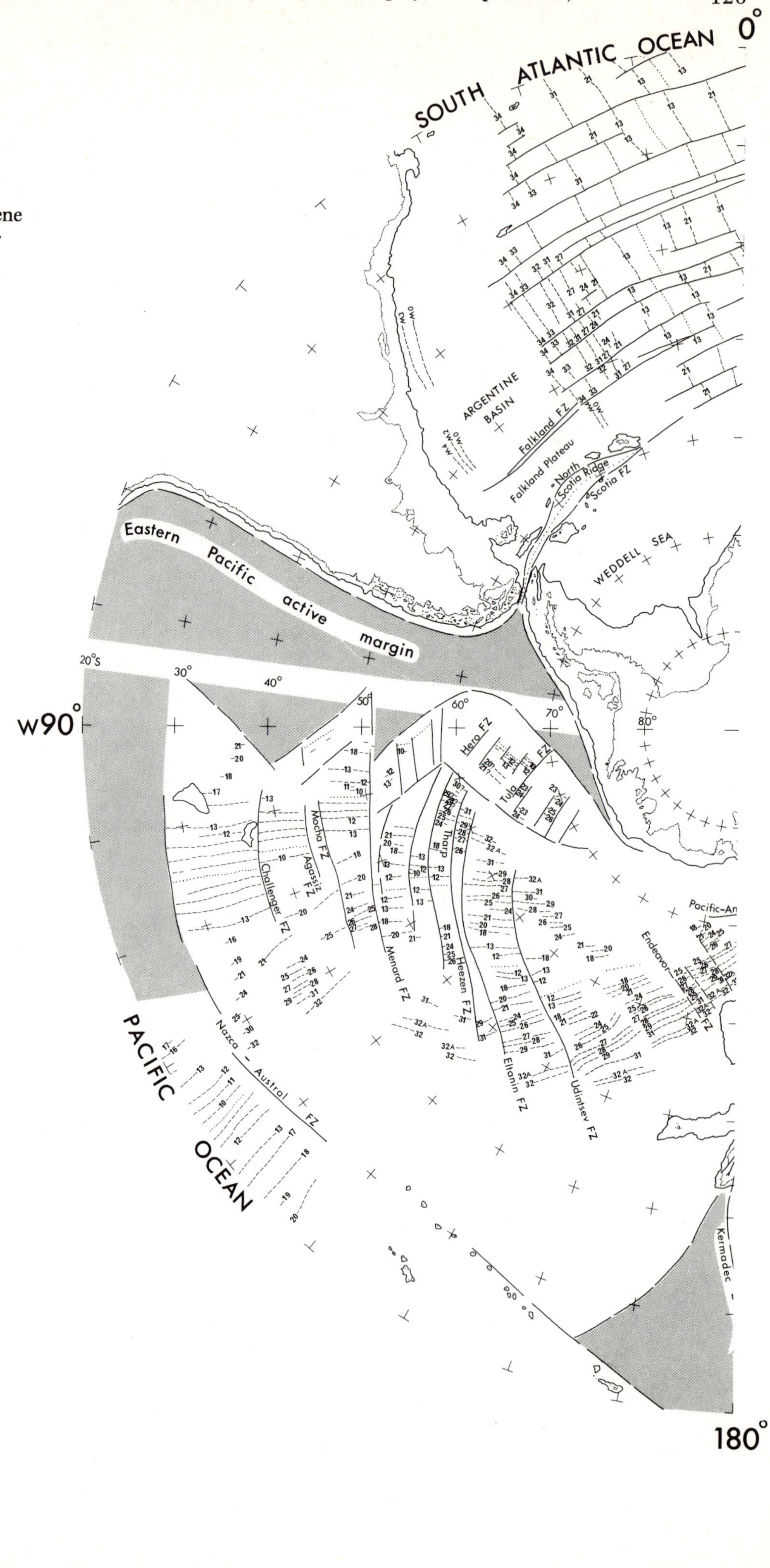

SOUTH ATLANTIC OCEAN 0°

ARGENTINE BASIN

Falkland FZ

Falkland Plateau

North Scotia Ridge

Scotia FZ

WEDDELL SEA

Eastern Pacific active margin

20°S 30° 40° 50° 60° 70° 80°

Hero FZ

Tula FZ

Tharp

Menard FZ

Heezen FZ

Eltanin FZ

Udintsev FZ

Pacific-An

Endeavor

W90°

Challenger FZ

Agassiz FZ

Mocha FZ

Nazca - Austral FZ

PACIFIC OCEAN

Kermadec

180°

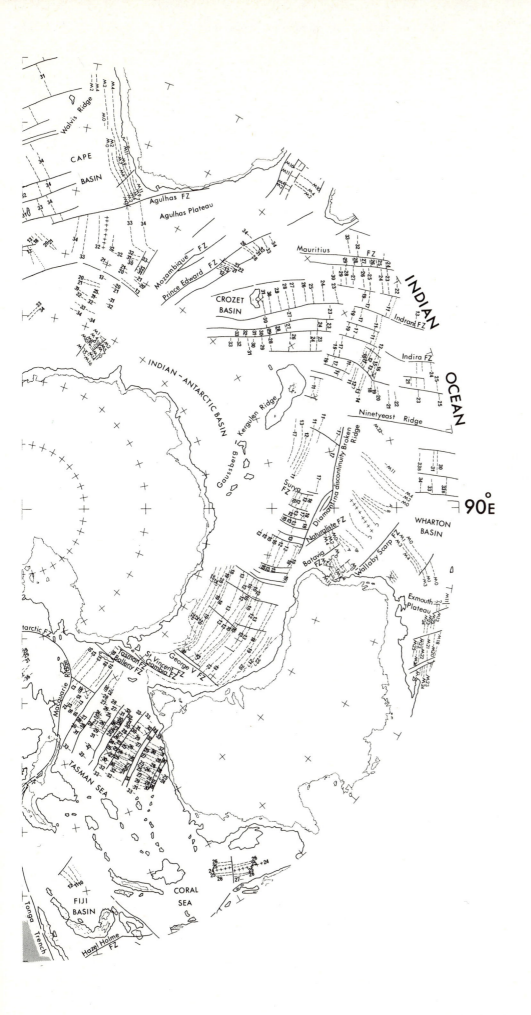

0°

Map 56

Anomaly 24 (56 Ma) Palaeocene
modern dimensions Earth

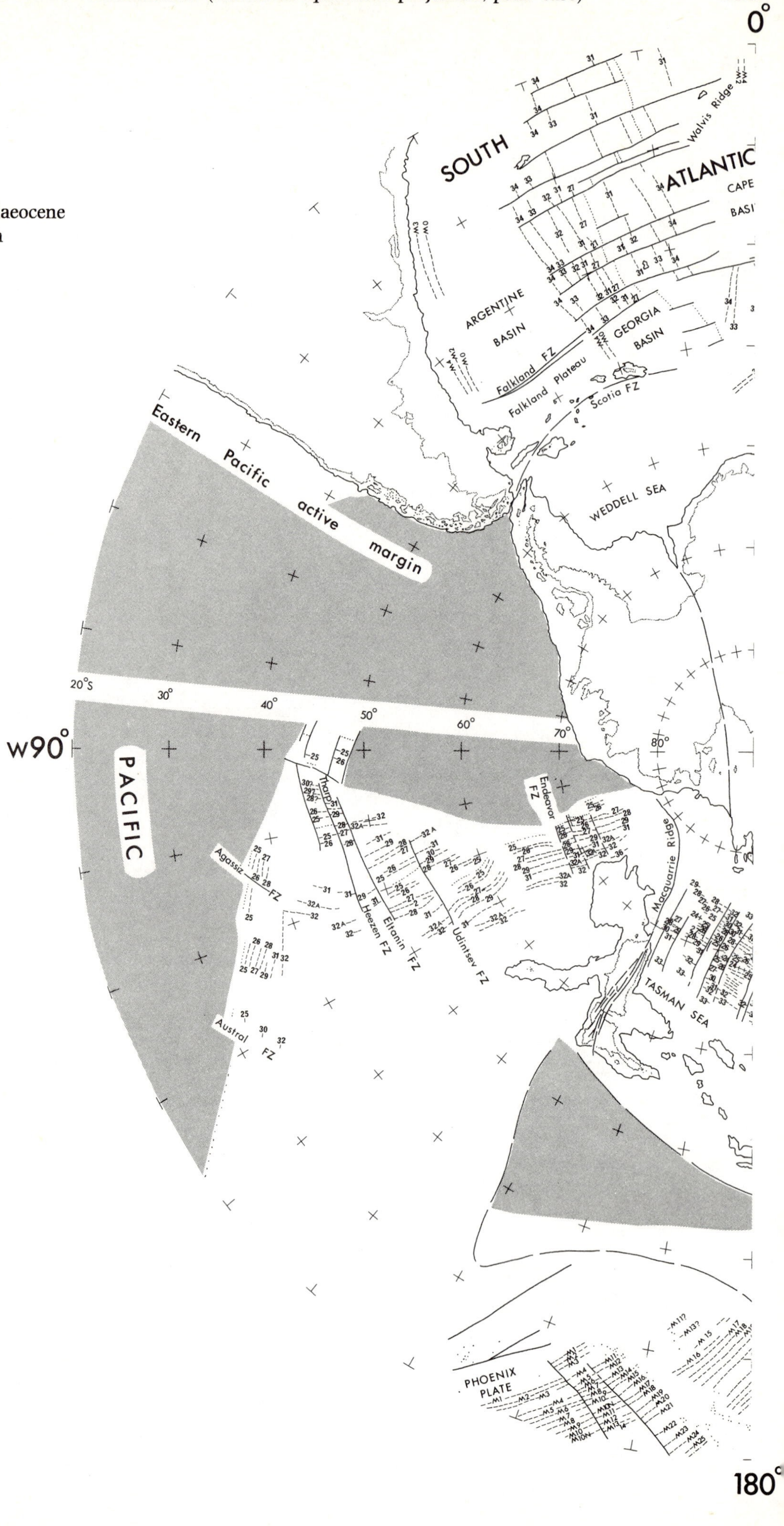

SOUTH

ATLANTIC

Walvis Ridge

CAPE

BASI

ARGENTINE
BASIN

GEORGIA

Falkland FZ

Falkland Plateau

BASIN

Scotia FZ

WEDDELL SEA

Eastern

Pacific

active

margin

20°S 30°

40°

50°

60°

70°

80°

W90°

PACIFIC

Endeavor FZ

Macquarie Ridge

Agassiz FZ

Tharp FZ

Heezen FZ

Eltanin FZ

Udintsev FZ

TASMAN SEA

Austral FZ

PHOENIX
PLATE

180°

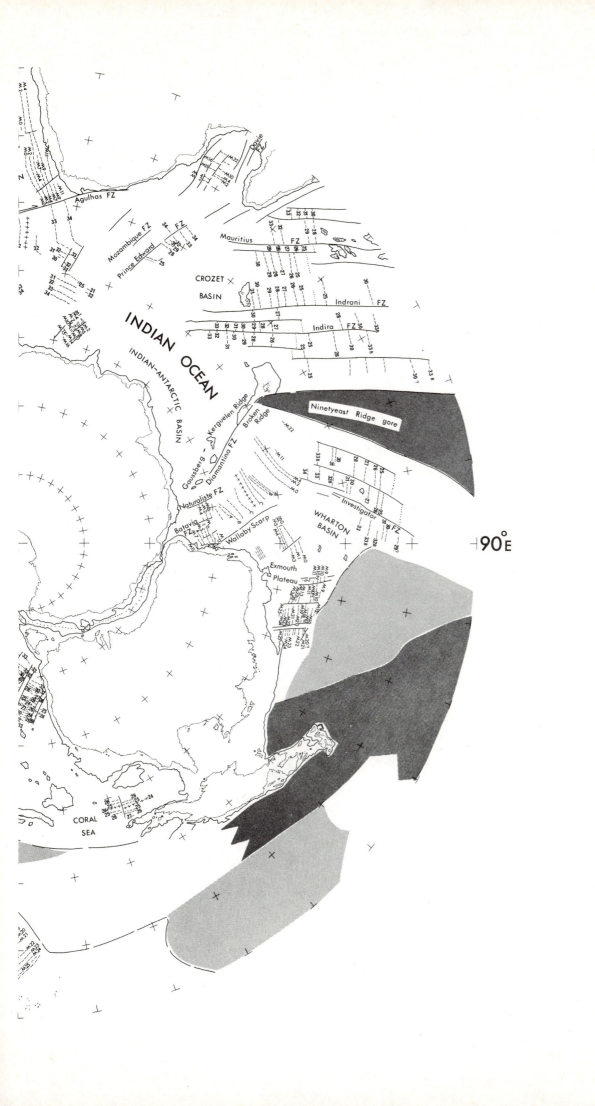

Map 57

Anomaly 24 (56 Ma) Palaeocene
ca 94% of modern diameter

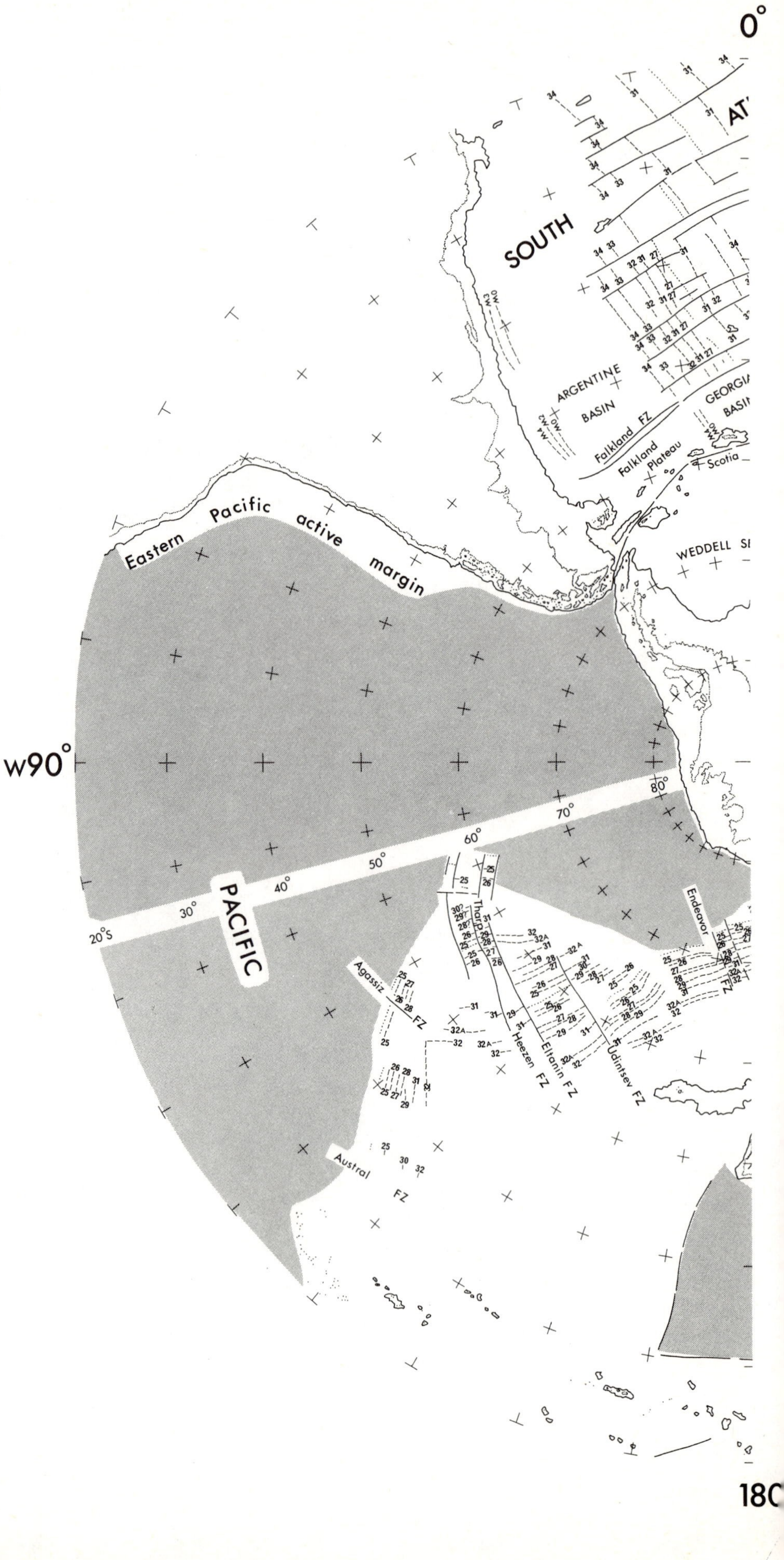

0°

SOUTH

AT

ARGENTINE

BASIN

Falkland FZ

GEORGIA

BASIN

Falkland Plateau

Scotia

WEDDELL S

Eastern   Pacific   active

margin

w90°

80°

70°

60°

50°

40°

30°

20°S

PACIFIC

Agassiz FZ

Tharp

Heezen FZ

Eltanin FZ

Udintsev FZ

Endeavor

FZ

Austral

FZ

180

Map 58

Turonian (90 Ma) upper Cretaceous
90% of modern diameter

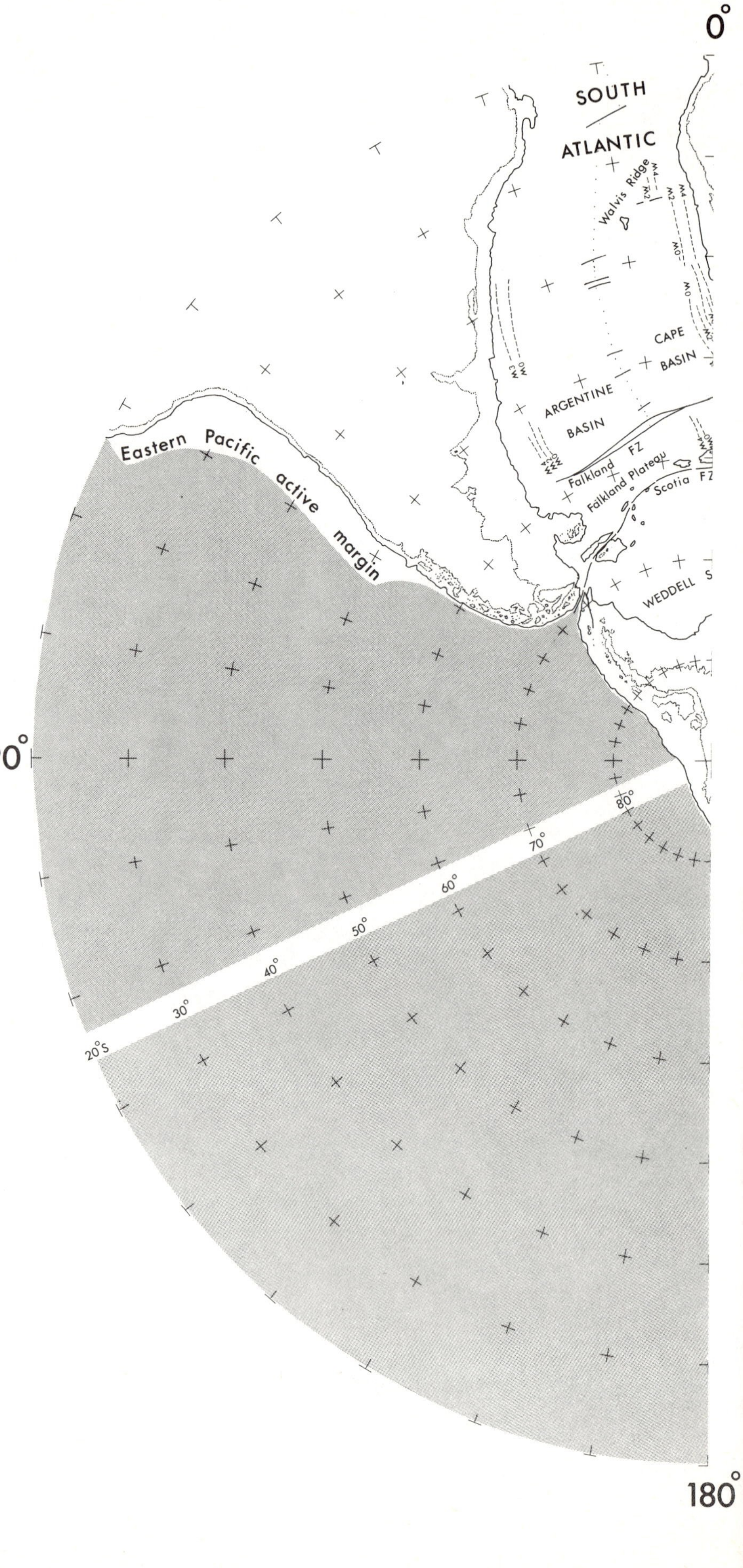

0°

SOUTH

ATLANTIC

Walvis Ridge

CAPE
BASIN

ARGENTINE
BASIN

Falkland FZ
Falkland Plateau
Scotia FZ

WEDDELL S

Eastern Pacific active margin

w90°

80°

70°

60°

50°

40°

30°

20°S

180°

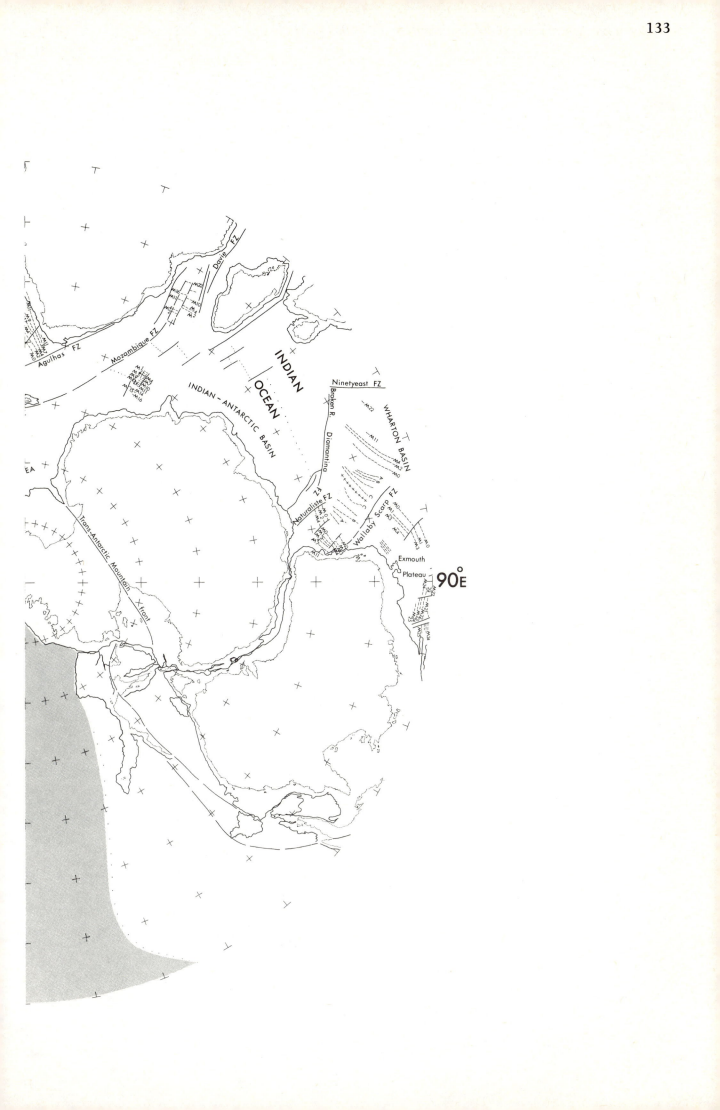

Map **59**

Anomaly M7 (120 Ma) Hauterivian
modern dimensions Earth

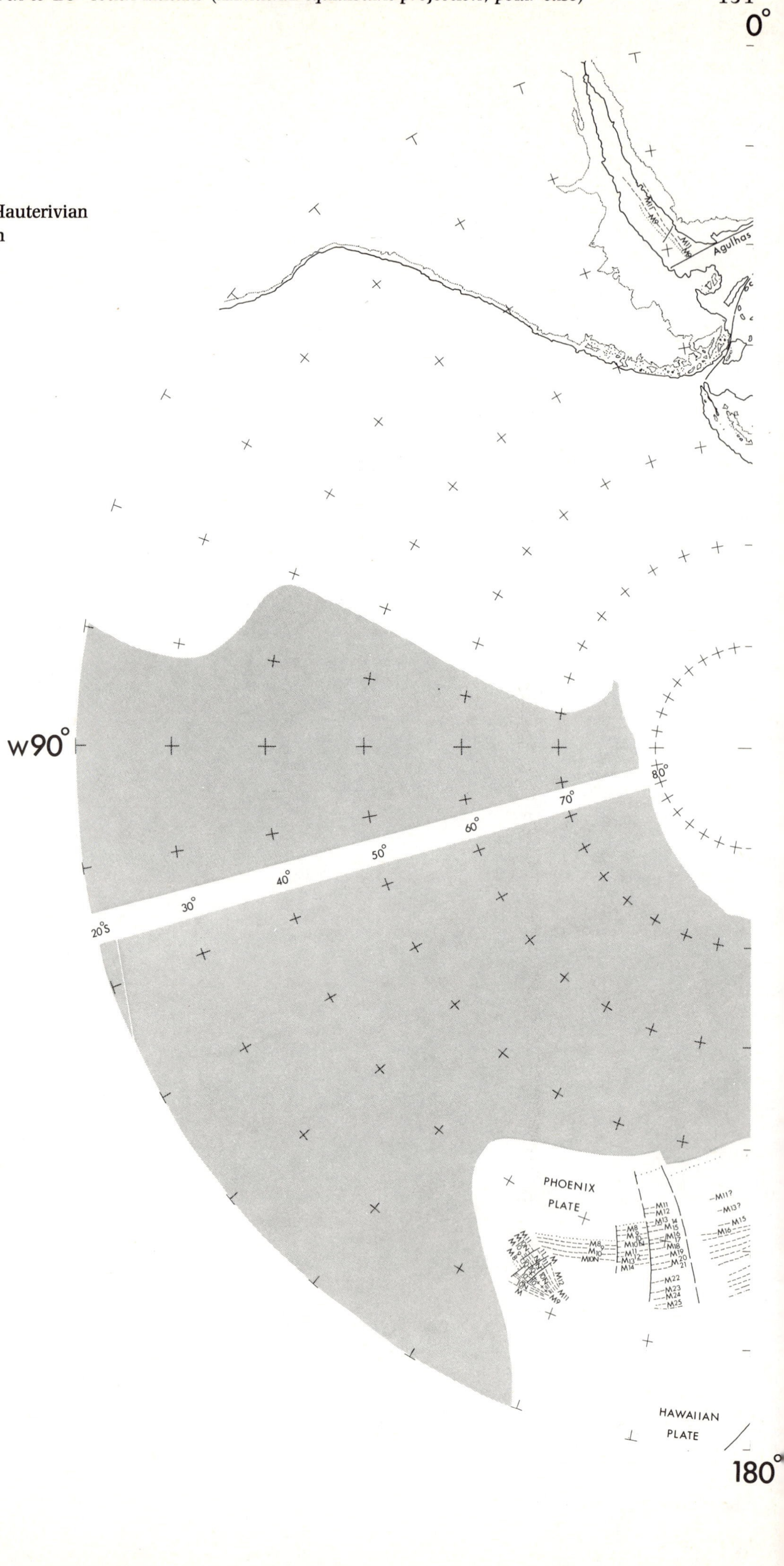

w90°

80°

70°

60°

50°

40°

30°

20°S

PHOENIX
PLATE

M11
M12
M13 14
M8 M15
M9 M16
M8 M10N M17
M10 M18
M12 M19
M13 M20
M14 M21

M11?
M13?

M15
M16

M22
M23
M24
M25

HAWAIIAN
PLATE

180°

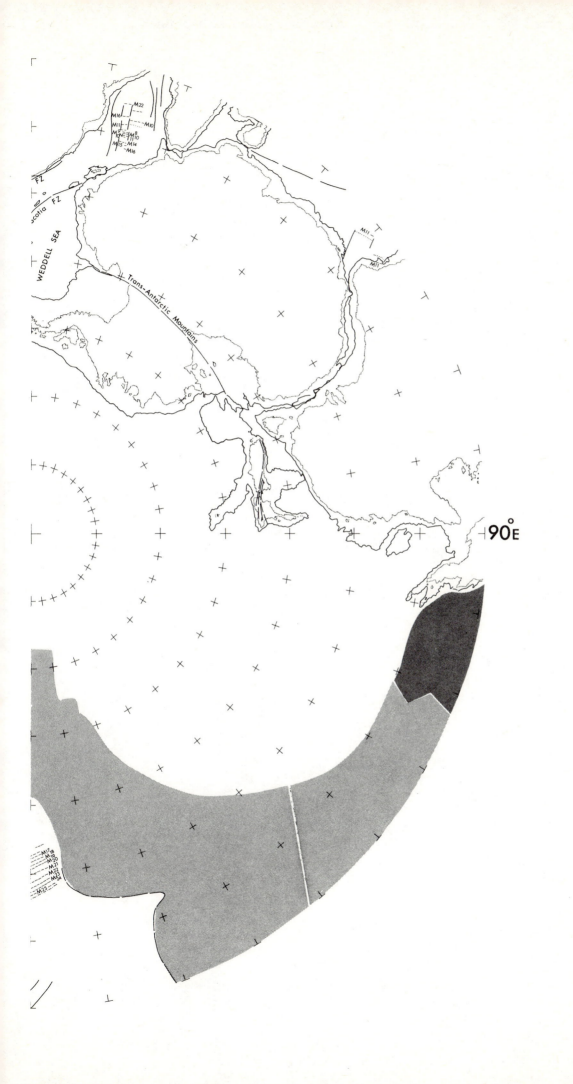

Map 60

Anomaly M7 (120 Ma) Hauterivian
ca 87% of modern diameter

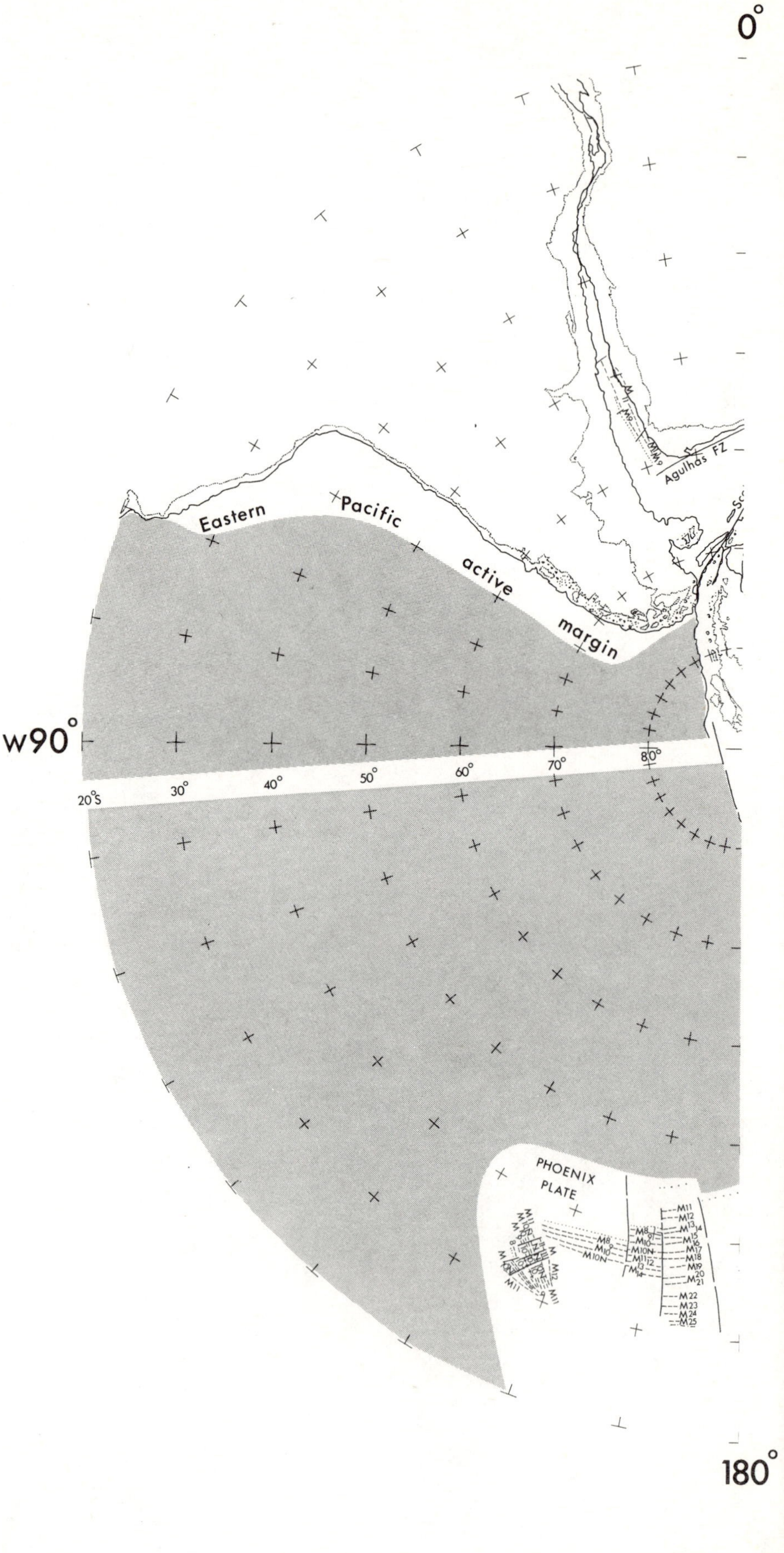

0°

Eastern   Pacific   active   margin

Agulhas FZ

w90°

20°S   30°   40°   50°   60°   70°   80°

PHOENIX PLATE

M11
M12
M13
M14
M8
M9
M10N
M11
M15
M16
M17
M18
M9
M20
M21
M22
M23
M24
M25

180°

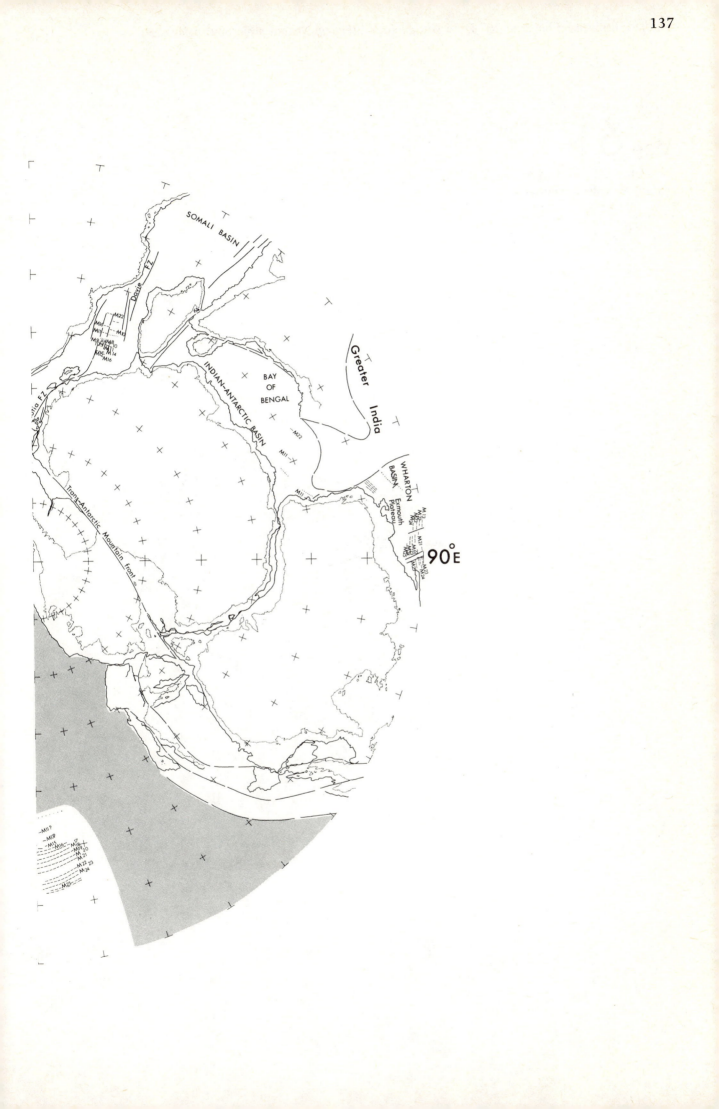

Map 61

Anomaly M23 (146 Ma) Oxfordian
ca 84% of modern diameter

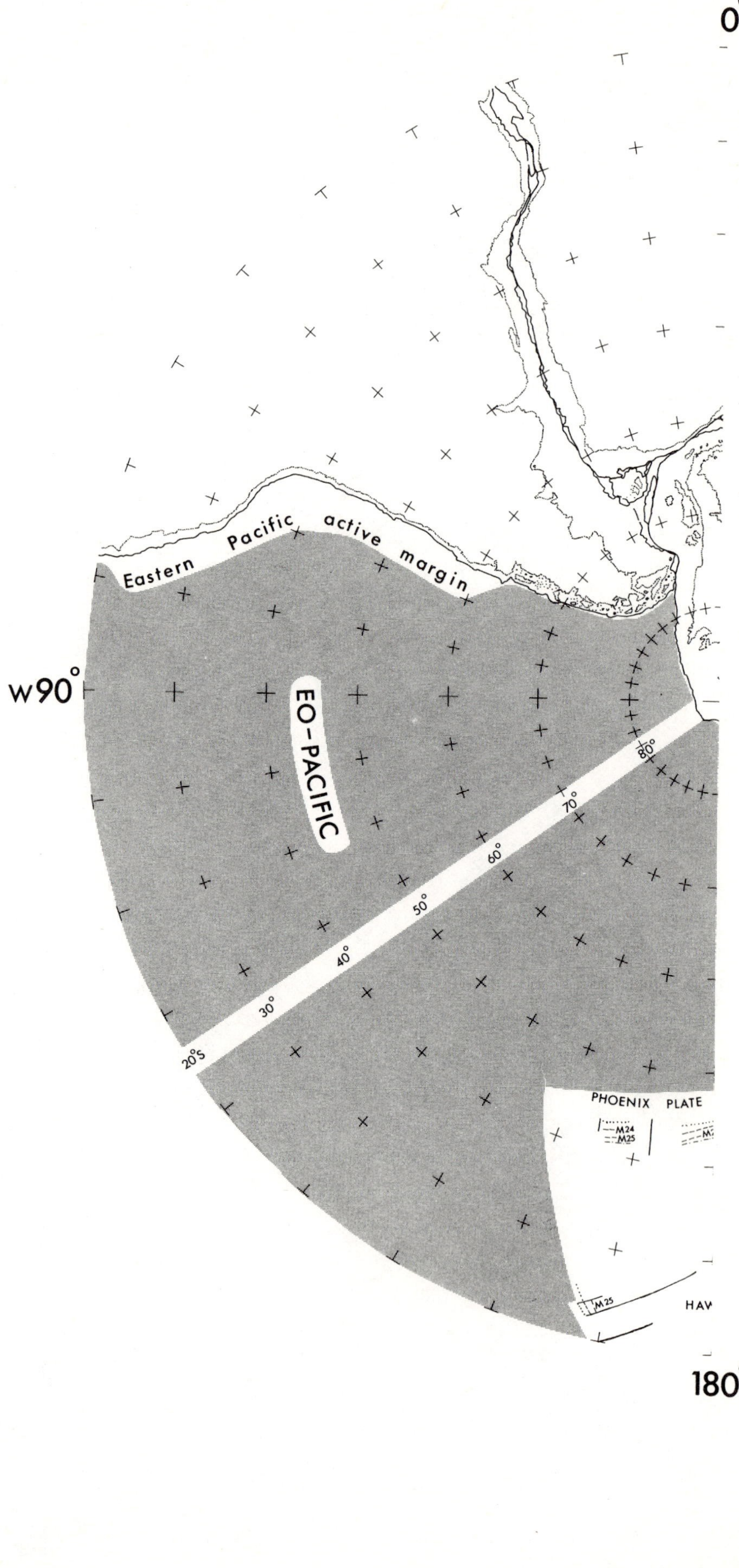

0°

w90°

Eastern Pacific active margin

EO-PACIFIC

80°

70°

60°

50°

40°

30°

20°S

PHOENIX PLATE

M24
M25

M25

HAV

180°

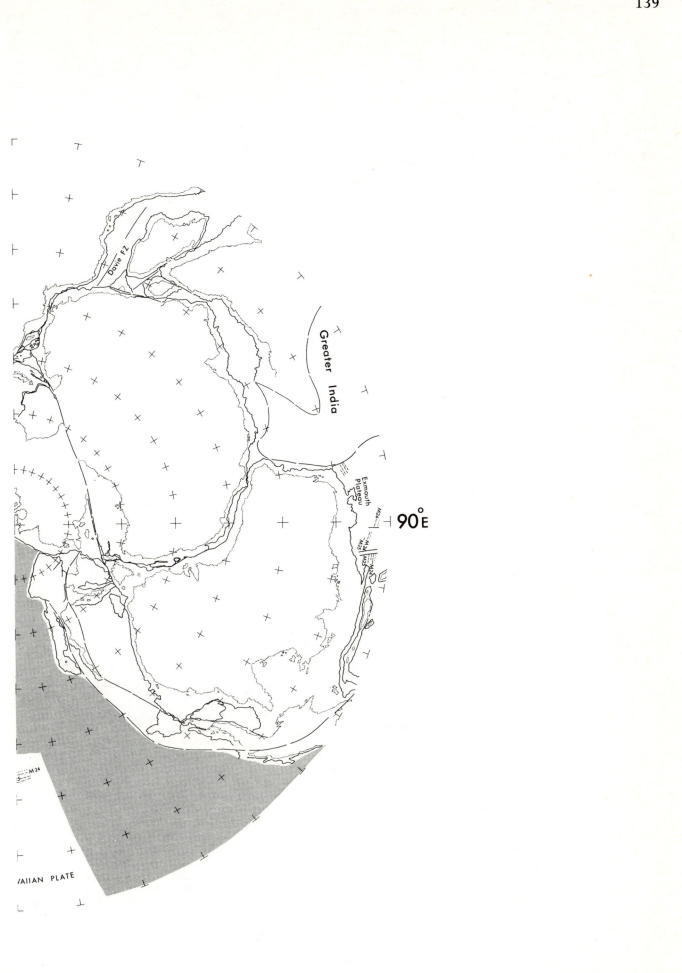

Dovie FZ

Greater India

Exmouth Plateau

M22
M24
M25

90°E

M24

/AIIAN PLATE

140
0°

Map 62

Pangaea (180–200 Ma) late Triassic to lower Jurassic
modern dimensions Earth

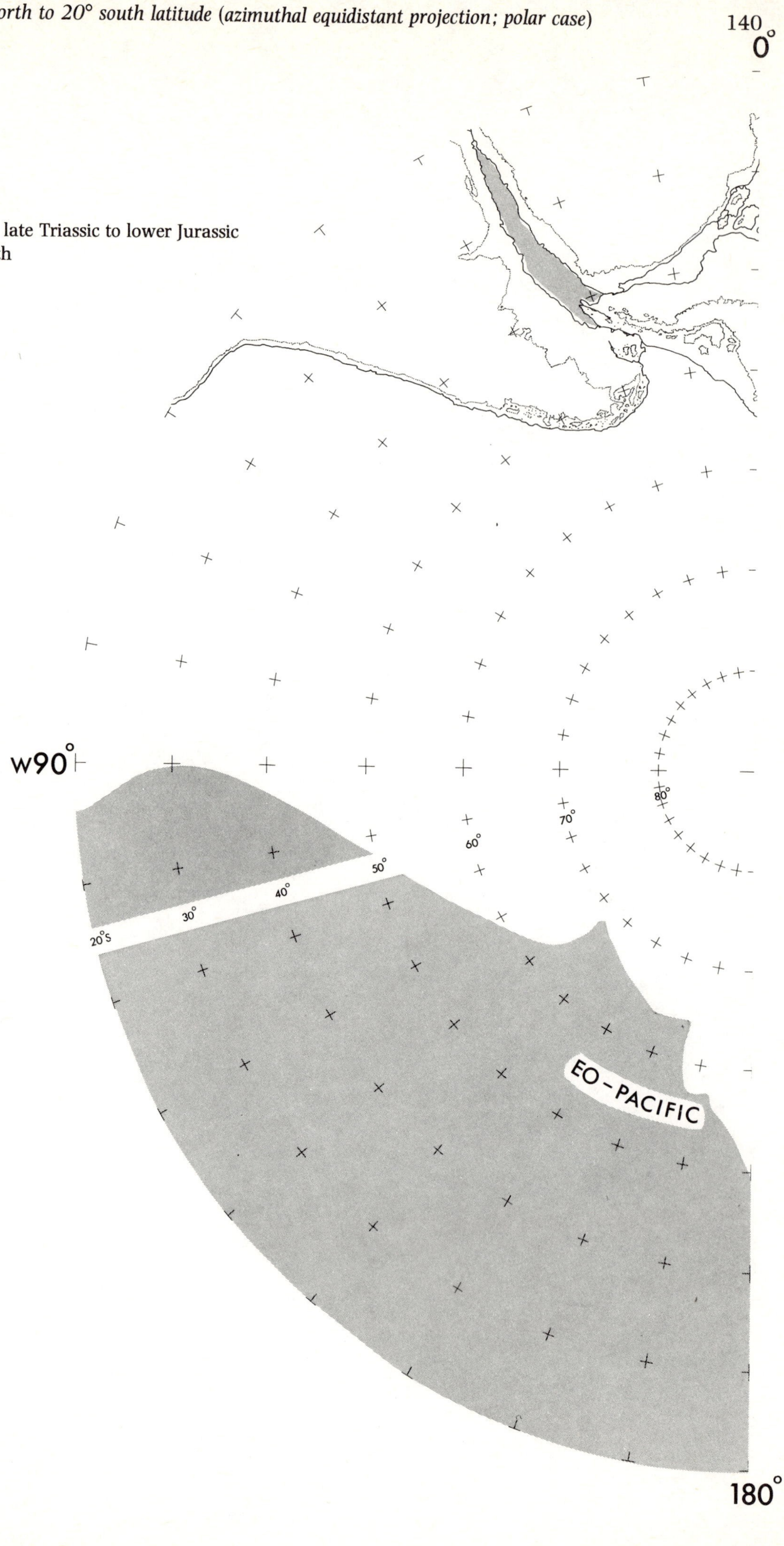

w90°

20°S
30°
40°
50°
60°
70°
80°

EO - PACIFIC

180°

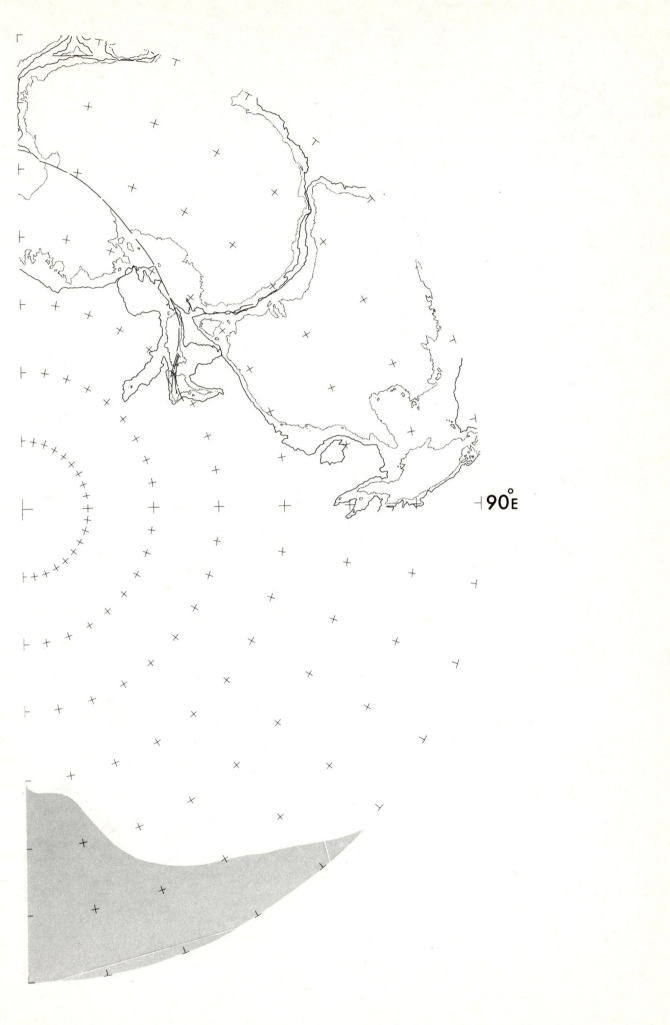

90°E

Map 63

Pangaea (180–200 Ma) late Triassic to lower Jurassic
80% of modern diameter

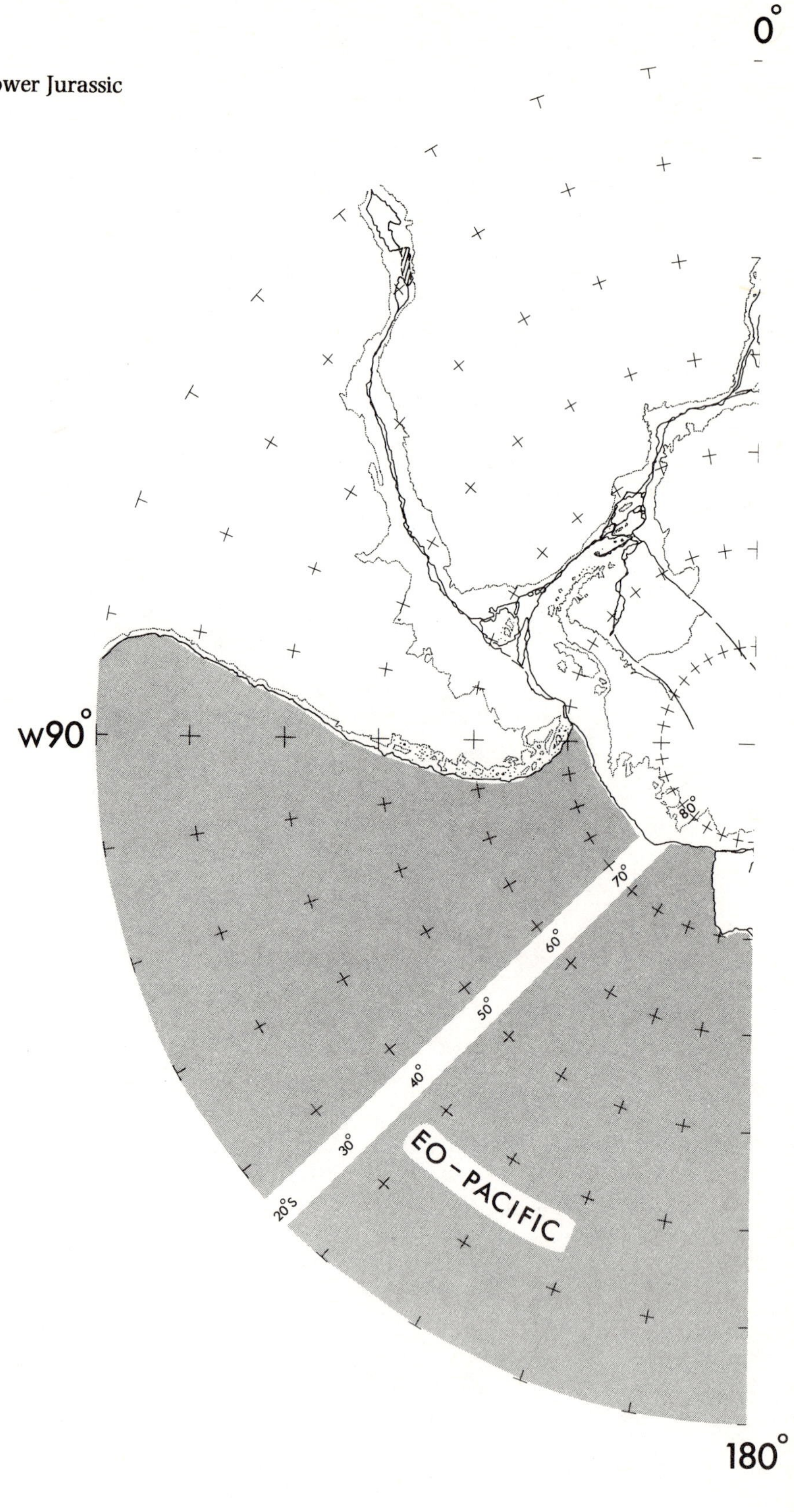

0°

w90°

EO - PACIFIC

180°

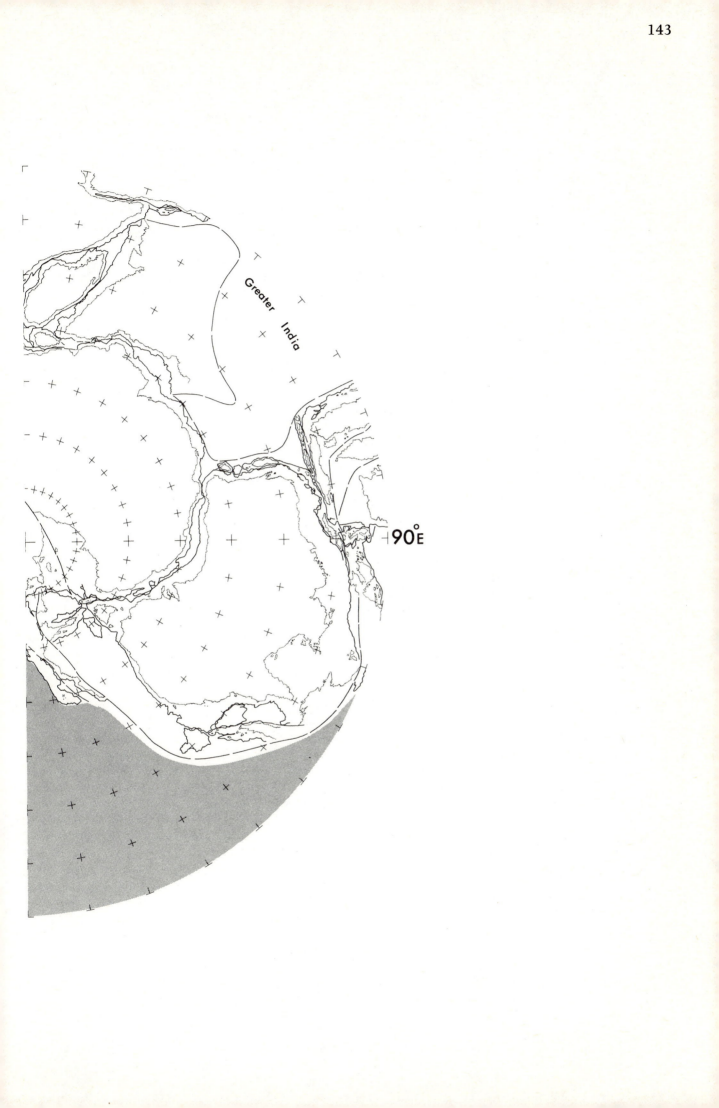

SECTION 7

World outline maps

MAPS 64–76

(Winkel's 'Tripel' projection with prime meridian at 10° east longitude)

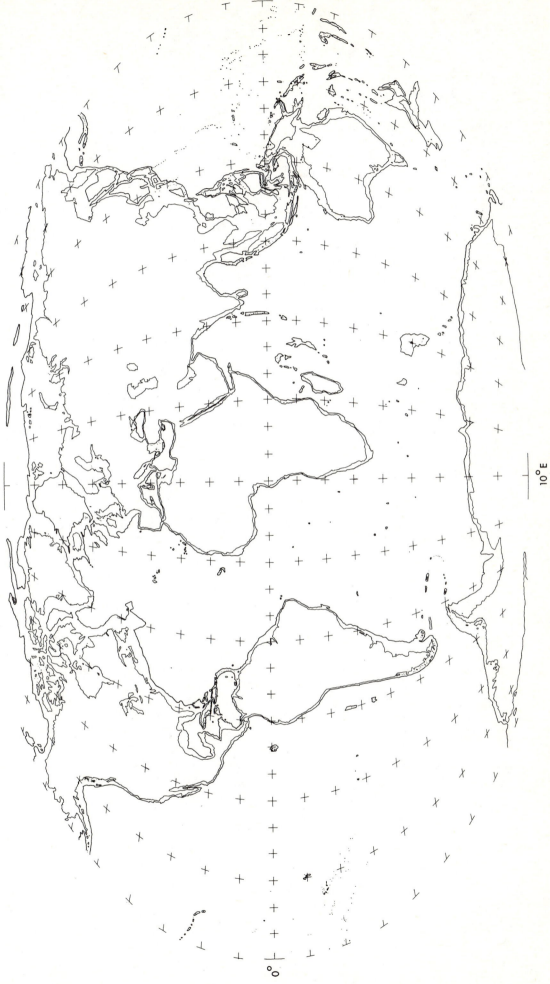

Map 64

Modern

10° E

0°

Map 65

Anomaly 9 (29 Ma) Oligocene
modern dimensions Earth

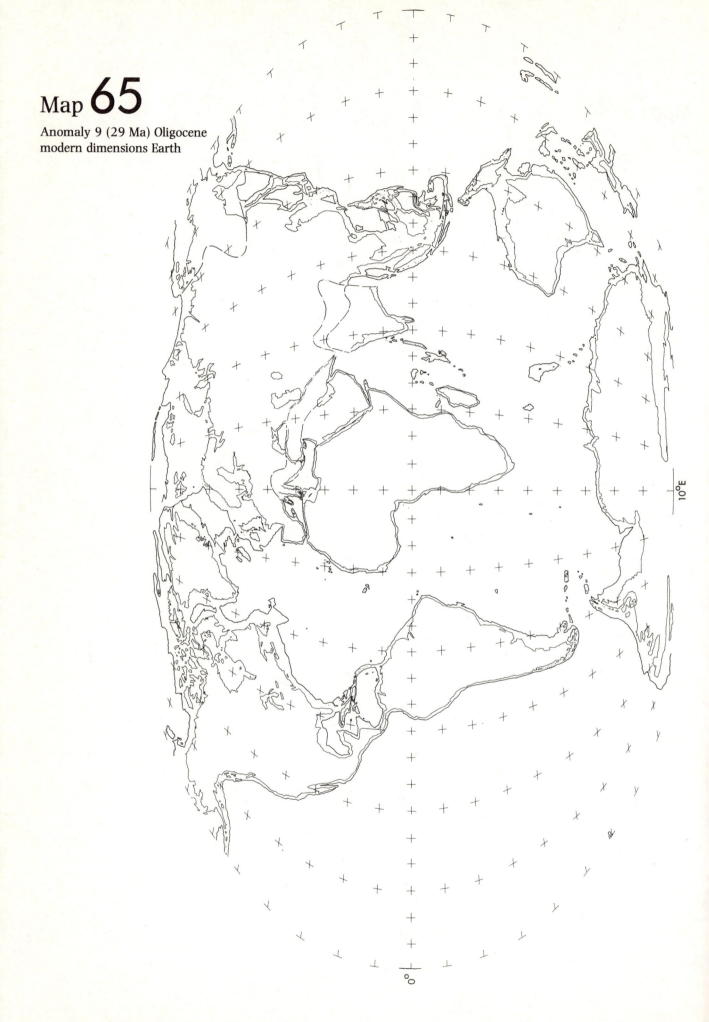

10°E

0°

Map 66

Anomaly 9 (29 Ma) Oligocene
ca 97% of modern diameter

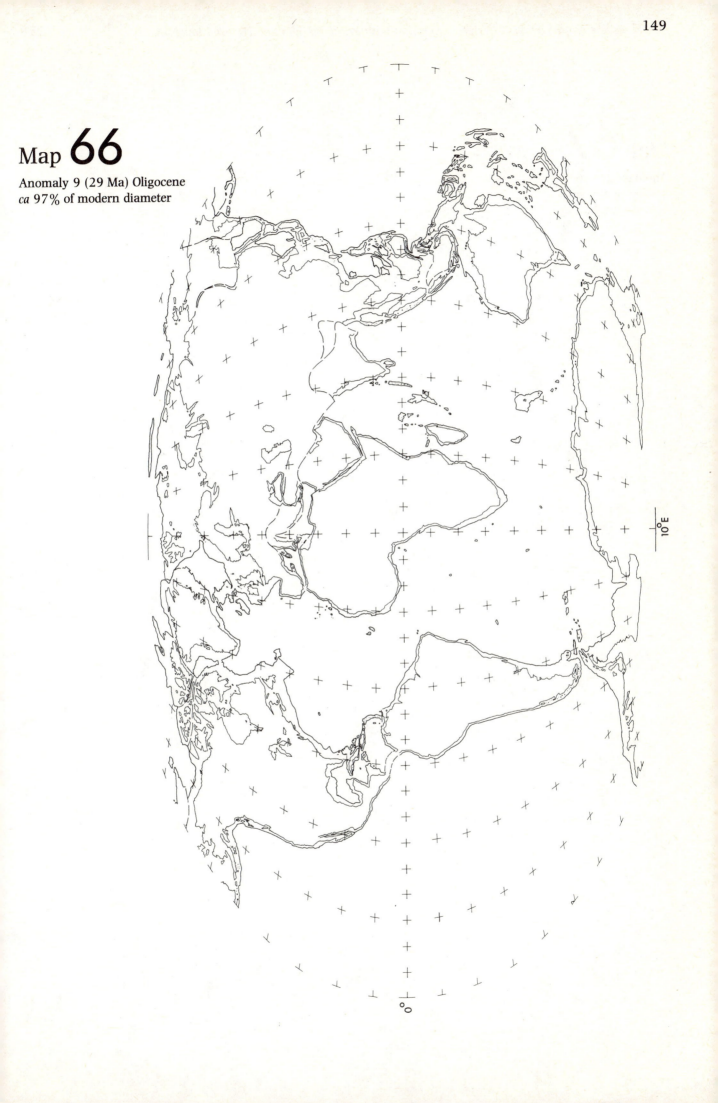

10°E

Map 67

Anomaly 24 (56 Ma)
Palaeocene modern
dimensions Earth

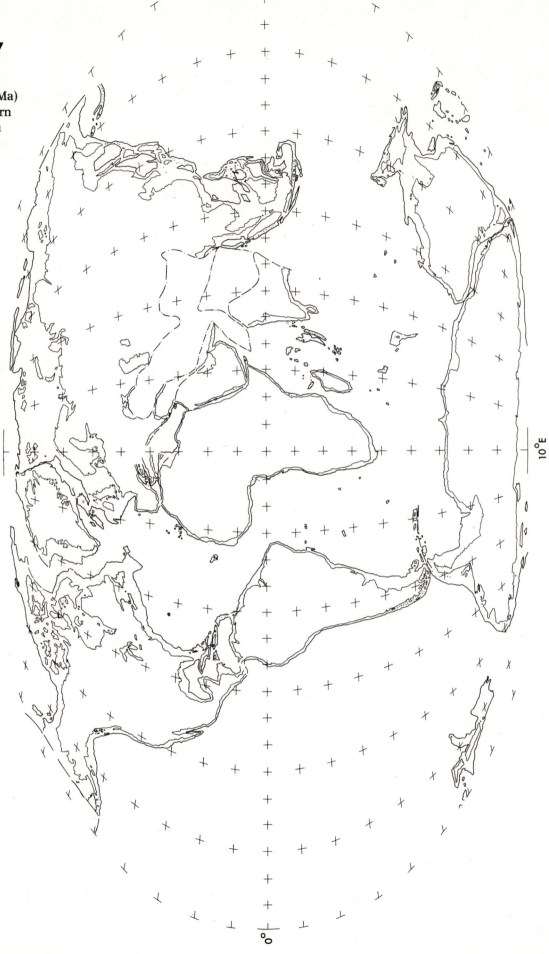

10°E

0°

Map 68

Anomaly 24 (56 Ma) Palaeocene
ca 94% of modern diameter

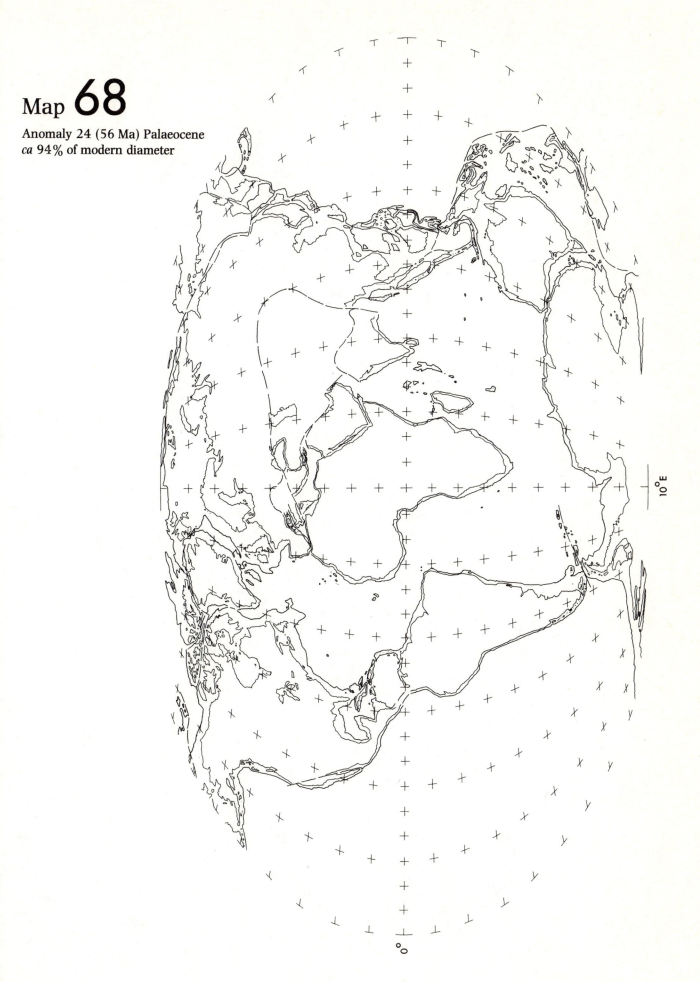

10°E

0°

Map 69

Turonian (90 Ma) upper
Cretaceous
modern dimensions Earth

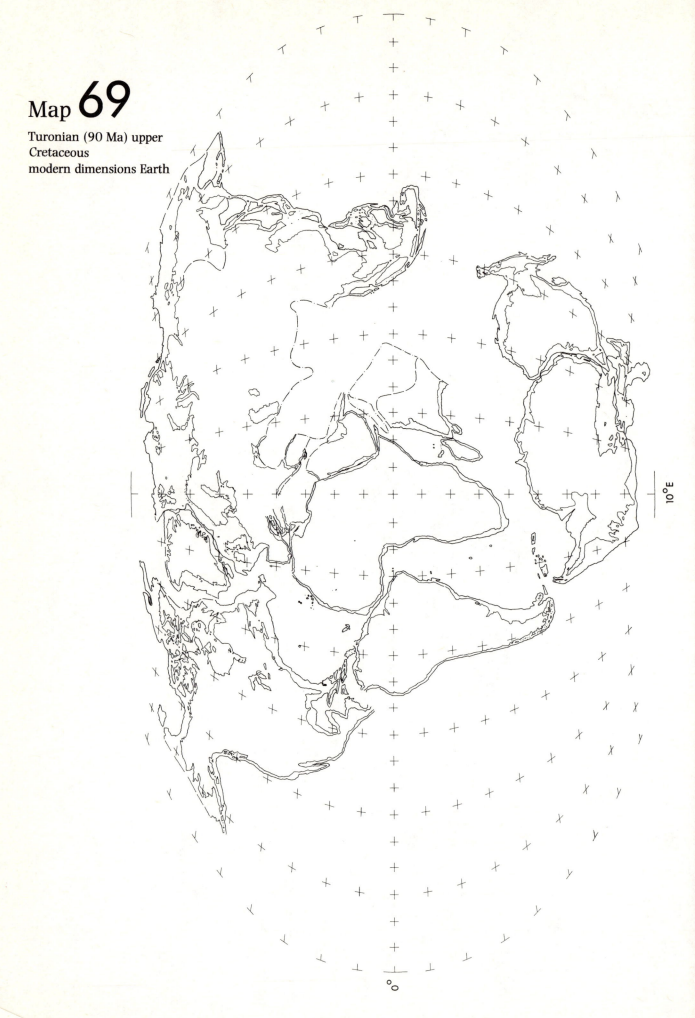

10°E

°0

Map 70

Turonian (90 Ma) upper Cretaceous
90% of modern diameter

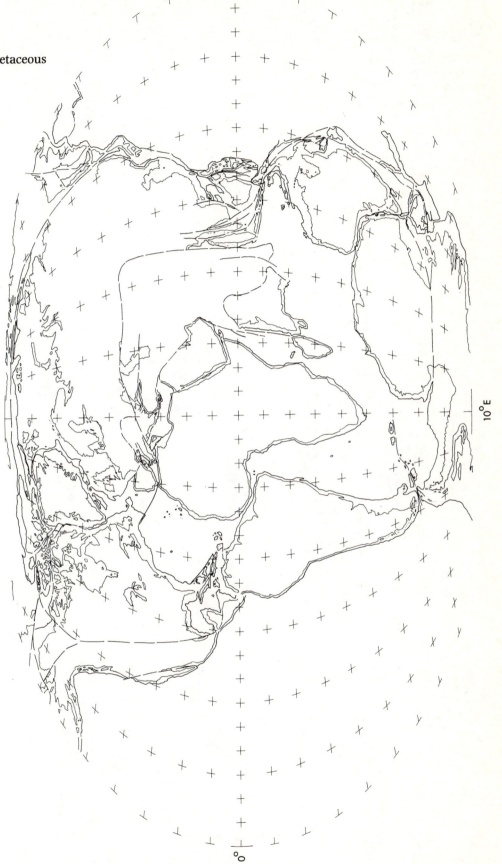

10°E

0°

Map 71

Anomaly M7 (120 Ma)
Hauterivian
modern dimensions Earth

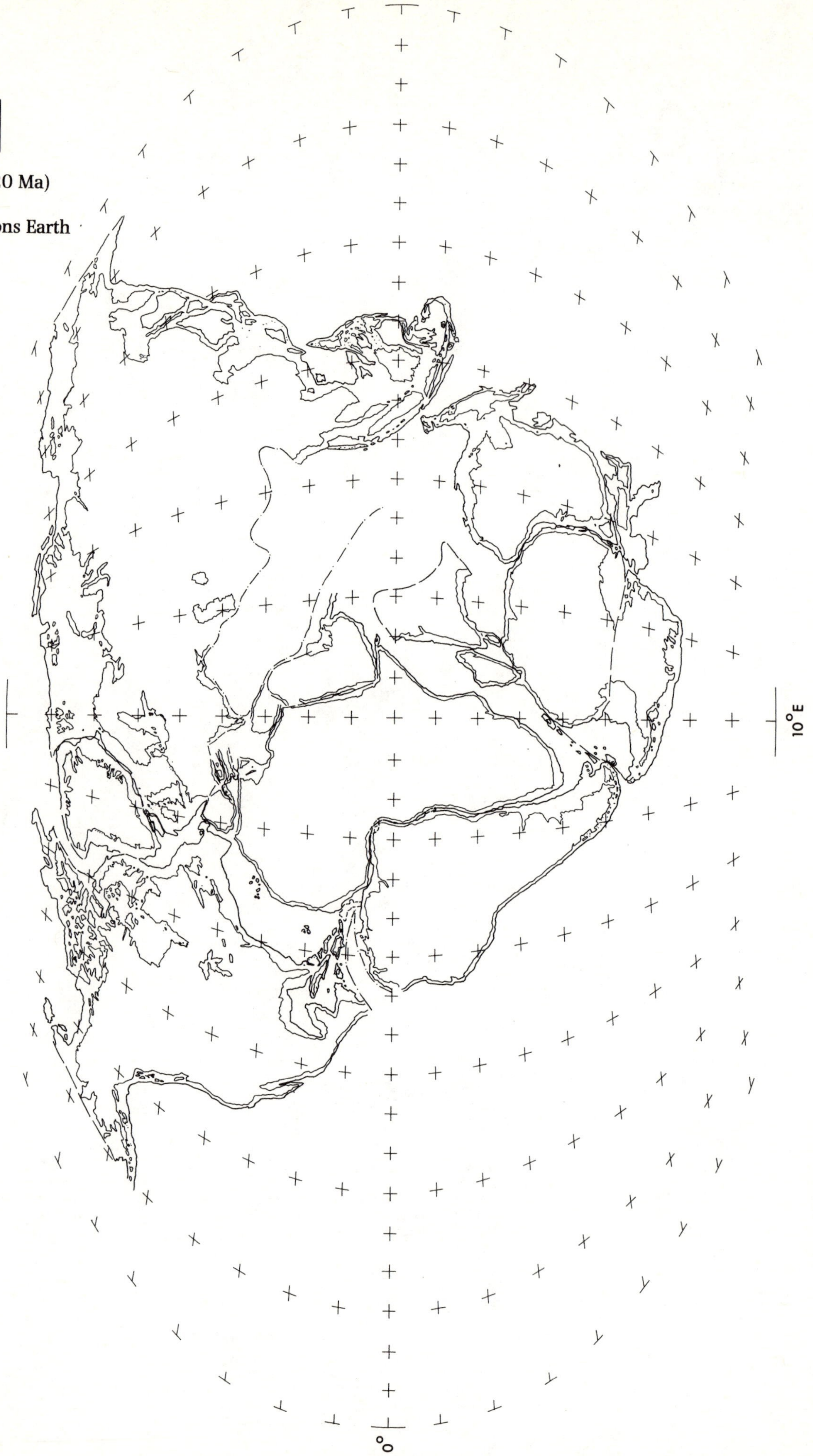

10°E

0°

Map 72

Anomaly M7 (120 Ma) Hauterivian
ca 87% of modern diameter

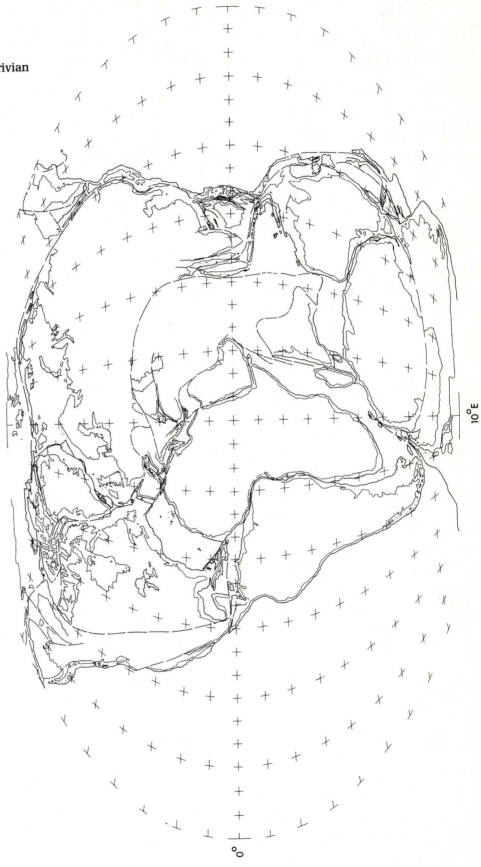

10°E

0°

Map 73

Anomaly M23 (146 Ma)
Oxfordian
modern dimensions Earth

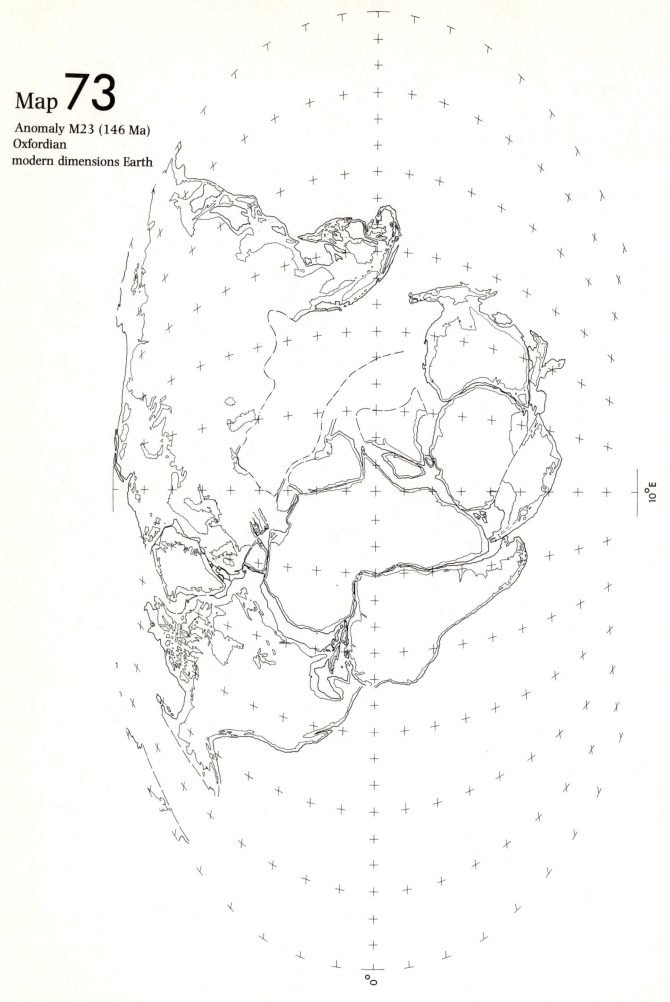

10°E

0°

Map 74

Anomaly M23 (146 Ma) Oxfordian
ca 84% of modern diameter

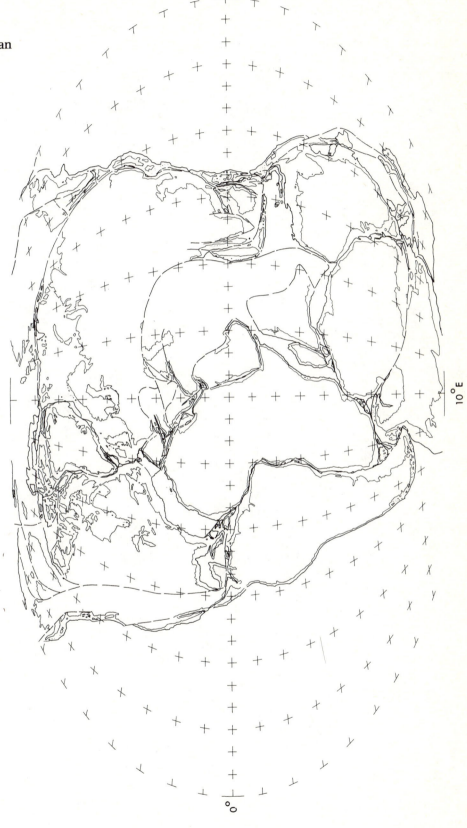

10° E

0°

Map 75

Pangaea (180–200 Ma) late
Triassic to lower Jurassic
modern dimensions Earth

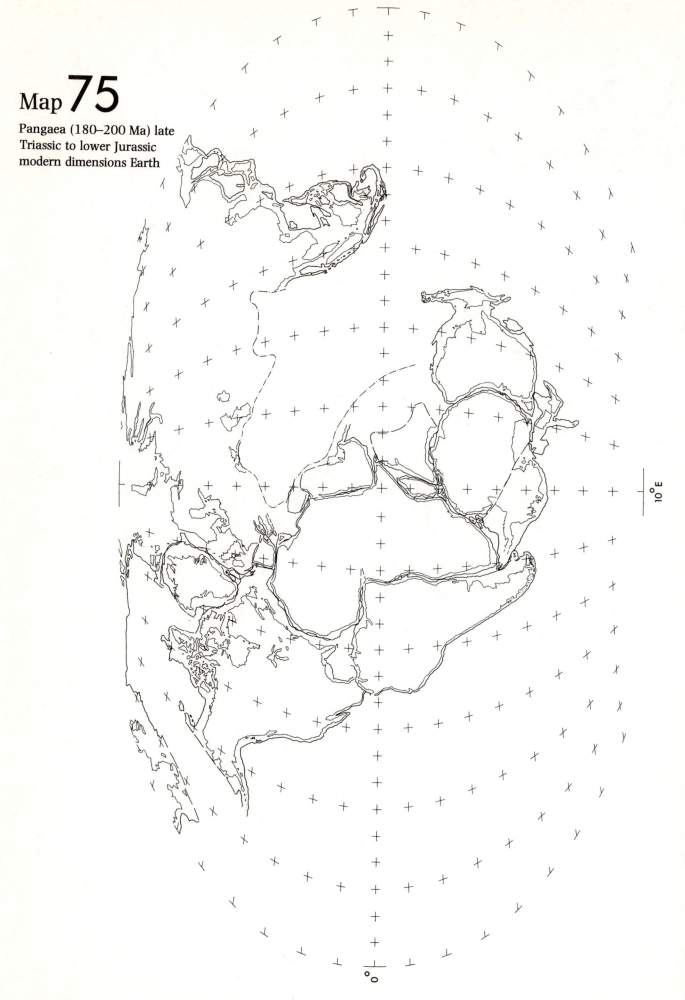

10° E

0°

Map 76

Pangaea (180–200 Ma) late Triassic
to lower Jurassic
80% of modern diameter

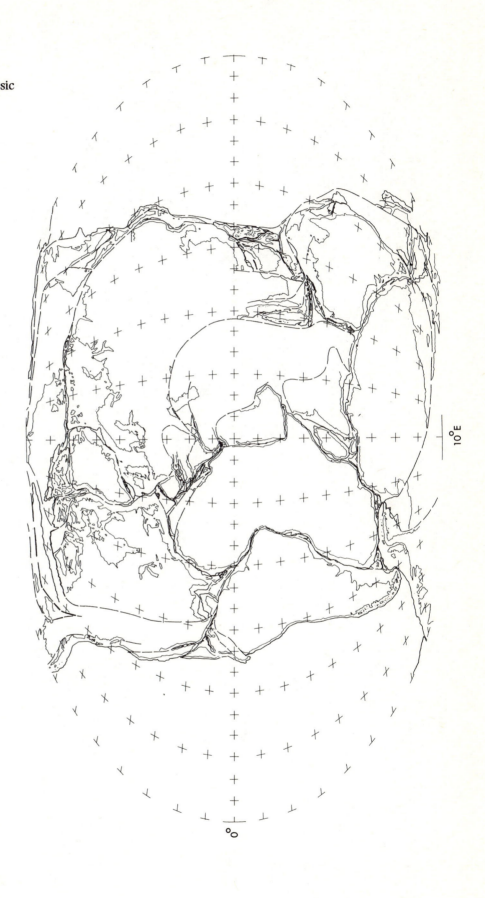

10°E

0°